A CONTINENT FOR SCIENCE

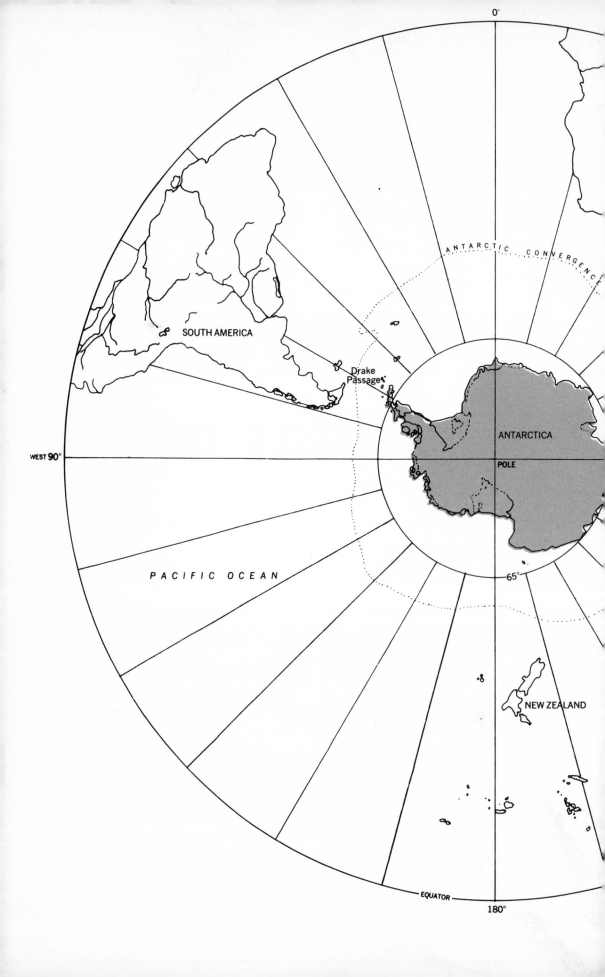

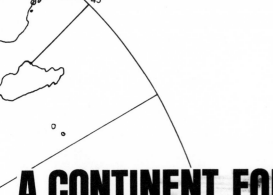

A CONTINENT FOR SCIENCE

THE ANTARCTIC ADVENTURE

RICHARD S. LEWIS

NEW YORK · THE VIKING PRESS

First published in 1965 by The Viking Press, Inc.
625 Madison Avenue, New York, N. Y. 10022

Published simultaneously in Canada by
The Macmillan Company of Canada Limited

Library of Congress catalog card number: 65-13204
Printed in U.S.A. by Halliday Lithograph Corp.

Second printing December 1966

ACKNOWLEDGMENTS

For use of certain photographs and maps, the author acknowledges with gratitude
the generous cooperation of the following institutions: The American Geographical
Society (map on title page); The National Science Foundation, Office of Antarctic
Research Program; The United States Navy; The Newberry Library, Chicago, Illinois
(for endpaper maps and map on page 9 from *Portugalia Monumenta Cartographica*,
1960); The New York Public Library (map on pages 6–7); *Transactions* of The
American Philosophical Society, Vol. 31, Part I, January 1939 (for map on page 11
from article by William Herbert Hobbs).

IN MEMORIAM

On the frontier of Antarctica triumph and death are inseparable companions. Many good men have found both there. One of them was Edward C. Thiel of Wausau, Wisconsin, a scientist of quiet sincerity and great competence. Dr. Thiel's work in Antarctica has enhanced the modern conception of that region and enriched the understanding of all men. To his memory *A Continent for Science* is respectfully dedicated.

FOREWORD

Antarctica is the most beautiful place I have ever seen. Yet it is an awesome land. There you can stand on a snow plateau at 8000 feet above sea level at 40 degrees below zero in a 40-knot wind and talk via short wave radio to friends in the northern United States who may tell you that the weather at home is colder than yours in Antarctica. There you can study forms of life adapted to the most severe conditions. There you may stand in the middle of an ice age—perhaps not unlike that which covered parts of North America. There you can see only a fraction of the earth which makes this continent, the rest being covered with ice averaging 7000 feet in thickness, and the character of what you cannot see must be determined by the scientific analysis of the data we have obtained. There you are in a virtual desert where the precipitation is less than an inch a year although the world's largest supply of fresh water (in the solid state) is underfoot. There you have, ready-made, a platform from which you can study to great advantage some of the electrical and magnetic phenomena which affect our communications. And as you think about our work you hope that this first of the great intensive environmental investigations of our time will continue on a wise and peaceful course. You begin to realize how little is known about Antarctica and you feel the challenge to learn about it.

Antarctica has been explored off and on for over a century, but it is only through the availability of modern transport—ships and aircraft—that the present comprehensive scientific effort has been made possible. The United States Antarctic Research Program, one of the many responsibilities of the National Science Foundation, is the crystallization of ideas and interests of scientists from many United States universities and government organizations. It is the result of scientists' working and planning cooperatively with their government. It requires sound planning in order to insure its success. Supply lines are long and the route to the south polar regions is treacherous and fraught with navigational hazards. Scientists working in the region must learn the art of survival in a desolate country where hidden crevasses, shifting ice floes, and bitter cold (temperatures often times drop below $-100°$ F) make just living a challenge.

There are eleven nations besides the United States with scientific interests in the Antarctic. To a large degree, the scientific programs designed and planned for the Antarctic reflect the flexibility or inflexibility of the systems which have generated them. We are all aware of the apparent success which controlled societies have achieved in short spurts through rigidly planned and organized programs. This poses a challenge to those of us in a free society who look upon "plans" with some degree of suspicion. To meet this challenge the United States has sought to develop an integrated research effort by building the program on the suggestions and thinking of outstanding scientists from all parts of the nation. National and international committees of scientists have contributed to the planning effort by making suggestions on the general guidelines which the United States research effort should follow. The individual scientists propose their own research projects. These are reviewed, selected, and combined to make a balanced United States program. In supporting this plan the United States Government channels its efforts to aid the individual scientist in his research. There is no dictation of, or restriction placed upon, individual ideas. Every effort is made to maintain an atmosphere in which creative thought can flourish. To administer such a program requires a special type of selflessness on the part of our staff in the Foundation. In these circumstances the scientist finds that his government has become his partner, providing a way for him to conduct scientific investigations of a relatively unknown continent. The success of this national program should be a major assurance to a free people that there are ways that man, working with his government on a cooperative basis, can meet the challenge of the rigidly planned programs of the regimented societies.

The work and the results of scientific research are like building a great wall, stone upon stone, part by part. All the stones and the mortar are needed. Most of the stones and nearly all the mortar will as the wall proceeds be unseen and may likely be remembered only by the particular men who did the work. The wall would not stand without each bit. Yet those who view the final result will likely be aware only of those pieces which make up the external façade. Stories which are written about long-range research programs generally point to the final results and seldom give the reader a sense of the way in which the knowledge evolved. Seldom have writers attempted to present some insight into the feelings which those who "built the wall" experienced as they worked.

The author of this volume has traveled to the Antarctic and has talked with many of our men who have worked in the field, in order to learn from them first-hand the joys and the fears of conducting scientific investigations in this remote region of the world. The author has sensed their sureties and their doubts. He has recorded the questions they asked, the challenges they faced. He has described their failures as well as their successes. He shows us that scientists are fallible, like everyone; that they often work against great odds, that they experience difficulties, frustrations, and failures from time to time, the same as everyone else. He describes how, through great persistence, they have succeeded in solving some of their problems. He has been able to capture the loneliness and desolation of this continent and project it in his writing so that as you read the book you feel that you are standing there "on the ice," facing the unknown. He vividly demonstrates man's complete dependence on his companions for survival in this wasteland of ice and snow.

The fact that the research work on this continent has served to spearhead political agreement could be another important result of these efforts. The nations working in Antarctica have drawn up a treaty setting aside this great continent for scientific and other peaceful uses. So, in addition, the work of the scientist becomes not only an adventure in research, but also an adventure in politics. The far-reaching possibilities of this unique international effort may be just as important to us as a political experiment as it is as a scientific program.

Following the entering into force of the Antarctic Treaty on June

23, 1961, representatives of the twelve governments have gathered together on three occasions to draft additional recommendations to ensure the smooth operation of the treaty and peaceful accord in Antarctica. Perhaps the most important of these recommendations to date is the set of Agreed Measures on Conservation. They are unusual because this may be the first time in history that man has ever seriously attempted any steps in conservation before it is too late.

As the population of the earth increases and as our ability to survive becomes more and more tenuous, man will find it necessary to adapt more and more the nature of this earth to his needs while preserving those features which are also so necessary for his survival. Our future on the earth may depend on our ability to cope with these unknown problems. The experiments in Antarctica, both in science and in politics, may be one of the necessary steps to this survival.

T. O. JONES

National Science Foundation
Washington, D.C.
January 5, 1965

CONTENTS

Foreword by Dr. Thomas O. Jones   vii

Preface   xvii

1. TERRA AUSTRALIS NONDAM COGNITA   3

Aristotle's Antarktikos · Mercator's king-sized land · The
Panhandle pointing north · Search for the south magnetic
pole · Landfall · Murray's hypothetical continent · "These
great southern solitudes"

2. THE SEVENTH CONTINENT   35

The polar race · Victory and death · Thin ice or thick · One
continent or two · The birth of the IGY · The Seven Cities
of the seventh continent

3. THE DEEP SEA OF ICE   72

Highway to Byrd Land · The ice thickens · Crossing the ice
bell · Assault on Victoria Land · "More ice than anyone
thought . . ."

4. RIDDLE OF THE LAND BENEATH THE ICE   95

A peripatetic penguin · The Grand Chasm · The trough
vanishes · Sea bottom versus continental land · Post-IGY begins

5. GONDWANALAND   125

The linkage of life • The natural history of Pangaea • Opponents of the supercontinent theory • Rivers of rock in the earth • Ice age upon ice age

6. THE VAULTS OF TIME   146

Sorge's law • The core explodes • Reading the oxygen calendar • Interglacials of McMurdo Sound • Rising beaches • The glowing rocks

7. MINUS 126.9 DEGREES FAHRENHEIT   167

Weather Central • Blackout • A ring of cyclones • Not one ice cap but two • The last clean milk • The heat sink • A warmer earth • Climate in an ice age • Arctic versus Antarctic • The last desert

8. THE DAWN CHORUS   199

How high is high? • Infernos of the sky • Do-it-yourself auroras • Whistler's father • Ions over the south pole • Forward scatter • The heavenly glow

9. ON THE FRINGE OF THE LIFELESS LATITUDES   224

Birds that walk like men • The private life of penguins • The "bugged" egg • Savaged by a penguin • Footprints in the snow • Bipolar birds • The fur-seal massacre • Contented true seals • Thar she blows! • Hijacking the seals • Antarctic fish story

10. THE ANTARCTICANS   255

Heroes with runny noses • The polar doctor book • Cold feet • Illusions on ice • The womanless world • White nights and black days • The outpost society • Antarcticans versus astronauts • A land without passports

Reference Notes   289

Index   293

PHOTOGRAPHS, MAPS, AND CHARTS

Mappemonde by Florianus, showing Terra Australis   6–7

Globe engraved in Paris, 1575   9

Map based on the Bransfield-Smith expedition   11

Map showing Antarctica in relation to other continents   19

Mount Erebus, photographed by Ponting in 1911   21

Hut Point, where Scott built his 1902 encampment (USN*)   29

Ice formations at junction of Ross Island and Ice Shelf (USN)   32

A penguin rookery, photographed by Ponting   34

Adélie penguins and chicks (USN)   36

The *Terra Nova,* taken from on board by Ponting   39

Headquarters of the *Terra Nova* expedition   41

January scene in Antarctica, showing midsummer ice floes   41

Scott's birthday celebration, 1911   42

Aerial view of the Beardmore Glacier (USN)   45

Ponies and dogs awaiting trial for Scott's polar expedition   46

Captain Lawrence Oates and companion at Cape Evans, 1911   46

Cross erected on Observation Hill in memory of Scott and his party (USN)   50

Two views of the bunkroom in Scott's hut, after restoration in 1961 (USN)   51

Sir Charles Wright in Antarctica in 1960 (USN)   53

Boulders in Wright Valley (NSF**)   55

Map of US IGY Antarctic Program expeditions   64

Construction crew at Little America Station   65

Seabees at Hallett Station (USN)   65

The icebreaker USS *Glacier* (USN)   66

Unloading cargo to be pulled to McMurdo Station (USN)   66–67

Night scene of main street at NAF McMurdo (USN)   67

Looking westward over McMurdo Station to Victoria Land   68–69

Dr. Harry Wexler at dedication for South Pole Station (USN)   70

Map of Bentley-Anderson traverse, Little America to Byrd Station, 1957   74

Sno-Cats poised to take off for a traverse (USN)   75

Seabees taking an ice coring in McMurdo Sound (USN)   77

Mount Erebus, overlooking the junction of the ice shelf with Ross Island   77

Icebreaker, tractor, and C-130 Hercules aircraft (USN)   84

Map of British crossing of Antarctica, 1957–1958   86

Ski-equipped Dakota airplane (USN)   88

Members of the 1960 traverse to the south pole   91

A Rolligon towed by a Sno-Cat (USN)   91

* Official United States Navy photo.
** National Science Foundation photo.

Map showing Russian traverses, 1957–1960; McCrary's traverse, 1960–1961; Van der Hoeven's Victoria Land traverse from McMurdo Sound, 1959–1960   93

Map showing Bentley-Anderson traverse, 1957–1958   99

An Adélie penguin (USN)   101

Diagram showing comparison of magnetic values with rock topography   102

Map showing Filchner Ice Shelf traverse, 1957–1958, and Ellsworth-Byrd traverse, 1958–1959   105

Profile showing formation of land under the ice, based on data from Filchner Ice Shelf traverse, 1957–1958   106

Scientists straining the waters of Lake Bonney (USN)   109

USS *Staten Island* (USN)   109

Profile showing formation of land under the ice, based on data from Ellsworth-Byrd traverse, 1958–1959   113

Map showing Bentley's traverse, Byrd Station to Horlick Mountains, 1958–1959   114

Map showing Ross Ice Shelf traverse, 1957–1958, and Victoria Land traverse, 1958–1959   115

Stephen Den Hartog, University of Wisconsin (NSF)   116

Dr. Thomas O. Jones, Sir Vivian Fuchs, and Admiral David A. Tyree at shaft leading to new Byrd Station (USN)   119

Dog team (USN)   121

Neptune P-2V flying over Beardmore Glacier (USN)   123

Diagram showing gradual disintegration of Pangaea   126

Geologists below north ridge of Mount Weaver (NSF)   131

George Doumani, Ohio State University (NSF)   133

Map showing projection of continents in the Mesozoic   142

Photographers on Observation Hill (USN)   145

USS *Glacier* (USN)   147

University of Wisconsin scientists boring through ice at Lake Vanda (NSF)   149

University of California scientist measuring temperature of a pond (NSF)   157

Rock formation in Victoria Land (NSF)   159

University of Wisconsin geologist inspects ground near Walcott Glacier (NSF)   163

USS *Edisto,* encrusted with ice (USN)   166

A windmill operating devices that record ground temperature (NSF)   171

Automatic weather station installation at foot of Beardmore Glacier (USN)   171

Meteorological readings being taken at McMurdo Station (USN)   171

Plane after arrival during a severe spring storm (USN)   172

The author at Williams Field   175

A caravan of tracked jeeps   175

Blowing snow storm at South Pole Station (USN)   177

High winds blowing snow and ice crystals (USN)   180

Iceberg containing portions of Little America III (USN) 186

Disabled helicopter on Mount Discovery (USN) 188

Sagging roof of building at Byrd Station (USN) 197

Main tunnel for new Byrd Station constructed in 1961–1962 197

Sir Charles Wright at Cape Royds (USN) 205

Aurora study tower at Byrd Station (USN) 205

Henry M. Morozumi adjusting an oscilloscope (USN) 209

National Bureau of Standards scientist ascending tower at Byrd Station (NSF) 219

Neutron monitors in laboratory at McMurdo Station (NSF) 221

Transmitting tower at McMurdo Station (NSF) 221

Marine biologist studying plankton (USN) 226

Emperor penguins and chicks at Cape Crozier (NSF) 229

Adélie penguins at Cape Adare (USN) 231

Adélie penguins at Cape Royds (USN) 232–33

A female skua (USN) 237

Weddell seals 239

A crabeater seal (USN) 241

A seal with her pup (USN) 243

Taking a milk sample from a Weddell seal (NSF) 246

Stanford University zoologist observing Sei whales (NSF) 246

Virologist preparing to investigate Weddell seals (NSF) 248

Some common animal species found at sea bottom of McMurdo Sound (USN) 249

Stanford University zoologist examining *Dissostichus mawsoni* (NSF) 251

Biologist probing in the Ross Sea (USN) 253

New Zealand mountain climber instructing American scientists (NSF) 257

Crew unloading supplies from a Hercules aircraft (USN) 257

Surveyors on Mount Crach (USN) 258

Captain Earland E. Hedblom (USN) 261

McMurdo Station galley (USN) 266

Chow hall at Williams Field (USN) 266

McMurdo Station during the dark austral winter (USN) 268

Base chapel at Williams Field (USN) 270

Navy chaplain with "adopted" penguin chick (USN) 270

Dentist working on a patient (USN) 270

Food supplies in tunnel at New Byrd Station (USN) 270

Signs at McMurdo Station and Williams Field (USN) 271

Radio ham shack at McMurdo Station (USN) 274

PREFACE

From Washington, D.C., it is 9214 miles to McMurdo Sound, Antarctica, the main base of the United States Antarctic Research Program. I have taken that trip twice on Military Air Transport Service via San Francisco, Honolulu, Nadi in the Fiji Islands, and Christchurch, New Zealand. As a safety measure, the seats in the aircraft are reversed so that they face the tail. But once airborne, I had no impression of riding backward, except into time.

On this flight to "where the pole Antarctic hath any elevation above the horizon," there is a transition in geologic time, or so it seems. I left an interglacial or postglacial period in North America on a Monday morning and arrived early Friday to encounter an ice age in Antarctica.

The first evidence of it appears about six hours out of New Zealand as the aircraft bears steadily southward on the final 2200 miles across the southern ocean. Table-topped icebergs float majestically below in the blue Pacific, some as large as Manhattan, and an occasional one of even greater size.

The ice pack begins south of the Antarctic convergence, a line of demarcation between two species of ocean, where the cold currents from Antarctica fall turbulently below the warmer waters of the northern seas. Streaked with ragged leads of water, the gray-white ice pack stretches 200 miles. Sometimes it heaves gently. Sometimes it lies flat upon the water like old silver.

Then open water appears—ice and blue water. This is the Ross Sea, the great embayment of the Antarctic continent on the west.

As the aircraft turns into the embayment, the slate-gray peaks of the Trans-Antarctic Mountains become visible. They rise through the snow, barren in the flaring sunlight. Beyond them there is a faint gleam of the ice sheet of Victoria Land, fading off to the west in an infinity of haze.

This scene has a familiar Alpine quality, but the scale is much larger. Pyramids of bare rock which resemble the Matterhorn alternate with lower, rounded domes covered with snow. In the distance are ranges nearly buried in snow, and these have a scalloped look.

For two hours the plane flies along a mountainous seacoast in the bright sunshine of the austral spring. It is nearly midnight, New Zealand time. At this time of the year, in November, the sun does not set. It is merely higher in the sky at noon than at midnight.

Ahead, a thin white line begins to grow on the horizon—what the mariners of old used to call the "blink" of ice. It is the Ross Ice Shelf, the earth's largest mass of floating ice, covering the narrowing southern half of the triangular Ross Sea. Now the scene has changed from an outsized Alpine landscape to a blue-white polar icescape. The snow reflects the bluish tint of the sky.

Vectored by ground radio into McMurdo Sound, an arm of the Ross Sea, the aircraft begins its descent under standard ground-control procedure. The "Fasten Seat Belt" and "No Smoking" signs are lit, just as though the plane were landing at Seattle or Milwaukee. Down goes the landing gear with a metallic shriek and a locking thump.

For the last two hours of the flight, the temperature of the cabin has been falling. All the passengers, including scientists and Navy personnel, have been struggling into layers of polar clothing, issued to us in New Zealand. Now, bundled in parkas, overpants, and rubber "thermal" boots, we peer out the windows at the ice age coming up to meet us.

The nature of this ice age in the modern world presents to the scientific mind one of the great mysteries of the earth. What does it mean in terms of the evolution and the future of the planet? In Antarctica today one sees the analogue of the ice ages of North America and Europe during the Pleistocene epoch, which comprises about a million years of the earth's recent geologic history. At least four different times during this epoch, North America, Europe, and

Western Siberia lay under ice sheets up to a mile thick. Each of the Pleistocene glaciations lasted for thousands of years. Between these periods were interglacial ages when the land was fair and open, covered with forests, prairies, and plains, as it is today. These times, too, lasted for thousands of years.

No one knows today whether we live at the end of this cycle or whether we are in another interglacial phase of it. If the latter is the case, our descendants will face another ice age, a catastrophe by present standards that would erase most of the civilization existing now.

We are no more than eleven thousand years away from the end of the last glaciation, called the Wisconsin. At its height, the ice covered most of North America, from the Arctic Ocean to the Ohio River, and most of Europe, to London and Berlin. The scientist can read the record of ice sheets in the morainal piles of stones left by the glaciers as they retreated across the land, and in the sediments at the bottom of the sea, but he cannot know the mechanics of these great sheets unless he investigates the Antarctic analogy.

Antarctica is a natural laboratory where theories about the causes and the evolution of ice ages can be tested. The Antarctic ice cap today is as large in area as any of the Pleistocene ice sheets at their maximum, and quite possibly larger. It covers a roughly circular area of about 5,500,000 square miles—more than the combined size of the United States and Europe. Its outer boundaries are the Atlantic, Pacific, and Indian Oceans. Beneath the ice cap lies the only "lost" continent man has ever found. Instead of being under water, like the fabled Atlantis or Mu, continental Antarctica is sunk beneath a sea of ice. Only its highest mountains are visible. Their geology provides abundant clues to the vast mineral wealth buried under the ice.

Breaking the circular trend of the ice cap are two major embayments, the Ross Sea on the west, as I have described, and the Weddell Sea on the Atlantic Ocean side of the continent. These indentations confer upon the continental ice cap the shape of an enormous bell, which appears to be ringing toward Africa and India. From its cone, opposite South America, extends a long, curved peninsula of ice-clad mountains that forms the handle of the bell.

A view of this enormous ice cap from space would reveal how it flows from the high plateaus of the interior, through the mountains in titanic glacial rivers, down to the coasts, where it extrudes across the embayments in floating shelves. It presents the powerful illusion that it is an immutable feature of the earth. We know it is not. The evidence of fossil tree trunks and seams of coal in the mountains shows that once this land, like our own, was fair and open to the sun. It was temperate, subtropical, and forested, like our own.

At some time in the past, the climate changed. This must have happened more than once. Snow began to fall. Ice built up in the mountains and flowed to the lower elevations under the pull of gravity. It drowned the land and whatever life may have existed there. Slowly the land sank downward under its growing load of ice.

The mystery of the ice cap is tied inexorably to the nature of this transition. How did it come about? The scientist has no answer. He can merely deduce that the transition occurred, from the evidence of temperate and subtropical fossil vegetation some hundreds of millions of years old. When he discovers the conditions that produced it, he will know something about the history and mechanism of climatic changes.

Only in our time has the scientist, supported by military logistics and government appropriations, begun a long-term investigation of the dynamics of environmental change. Man's future on this planet may depend on his ability to understand and modify these forces. That is what the investigation of Antarctica means in the modern world—and what it has always meant. It is a very old investigation indeed.

A CONTINENT FOR SCIENCE

Beyond this flood a frozen continent

Lies dark and wild, beat with perpetual storms

Of whirlwind and dire hail, which on firm land

Thaws now, but gathers heap; and ruin seems

Of ancient pile; all else deep snow and ice. . . .

—John Milton, *Paradise Lost* (1667)

TERRA AUSTRALIS

NONDAM COGNITA

The investigation of Antarctica has challenged civilized men for twenty-five hundred years. Long before Mediterranean mariners ventured to the equator, the existence of a southern continent was assumed by ancient Greek scholars, who wrote about the land of the Antichthones, those who lived on the other side of the earth. The land itself was called Antichthon or Antarktikos, the antipode of the cold region in the north which lay under the constellation Arktos, the Bear. The concept was not a mystical one, nor did it have a legendary origin. It was a projection in geographic terms of man's awareness of bilateral symmetry in himself and in nature. As early as the sixth century B.C., Pythagoras taught that the world was a sphere. In order to confer equilibrium upon a spherical earth, his followers postulated the existence of large masses of land in an unknown southern hemisphere to balance those which formed the inhabited world in the north.

Aristotle's Antarktikos

Measuring the size of the terrestrial sphere was a fascinating problem which occupied mathematicians, geographers, and philosophers of the ancient world for centuries. An early estimate, ascribed to Eudoxus of Cnidos and Callippos of Cyzicos, was that the earth had

3

a circumference of 400,000 stadia, or about 40,000 miles. This figure was considered reasonable by Aristotle. He pointed out that the spherical shape of the earth should be obvious to everyone since, as one travels a short distance from north to south, new stars appear over the southern horizon and familiar ones in the northern sky disappear. Aristotle also argued that the existence of a southern continent was a logical necessity to preserve the balance and symmetry of the world. In its southern extremity, such a land might be cold, barren, and icy.

By Aristotle's time the Greeks were aware of cold lands near the Arctic Circle. Phoenician ships had been there and, in the fourth century B.C., Pytheas of Massilia (now Marseilles) discovered for the Greeks the island of Britain. He sailed beyond it to a land called Thule.* Greek merchant ships which followed Pytheas's track to trade for tin in the northern islands occasionally were blown far off course by fierce winter gales. Sailors saw frightening mountains of ice adrift in the sea. Through swirling fogs, they glimpsed high coasts covered with snow.

If such a region existed in the north, the rule of symmetry demanded its counterpart in the south. So was born the concept of Antichthon, Antarktikos, Terra Australis, Antarctica.

As mathematicians developed skill in determining latitude, they reduced earlier estimates of the size of the earth. A remarkably accurate estimate was made by the Alexandrian geographer Eratosthenes in the third century B.C. His point of departure for the measurement of the earth was the Egyptian town of Syene, the modern Aswan, at the first cataract of the Nile, 5000 stadia south of Alexandria. There the sun casts no shadow at noon on the day of the summer solstice. By measuring the angle of the shadow cast by the sun at this precise time in Alexandria, Eratosthenes calculated that the angular distance between Syene and Alexandria was 7 degrees, 12 minutes of arc, or 1/50 of a circle.** Since the distance between the two points was 5000 stadia, the earth's circumference was 250,000 stadia, or about 25,000 miles, amazingly close to the modern figure.

Posidonius, a Greek philosopher of the first century B.C., who figured the circumference at 24,000 miles—shorter than the actual value—made another contribution to practical geography. If you sailed due west from Gades (Cadiz) in Spain, he said, you would

* Probably one of the Shetland Islands, according to John Boyd Thatcher in *The Continent of America* (1896).
** The shadow was measured in a gnomon, a wooden bowl with a stick standing upright in the center.

reach India in 7000 miles. Thus, this idea was 1400 years old when Columbus presented it to Queen Isabella.

In Spain, Pomponius Mela, a geographer of the first century A.D., divided the planet into northern and southern hemispheres. He described five zones of climate: two frigid zones, a torrid zone, and two temperate zones, which were habitable. Mela could not doubt the existence of an inhabited counterpart of the northern world. Antichthones must live in a southern temperate zone, even if they were unknown and unreachable because of a zone of fiery heat between the hemispheres.

Reason alone dictated such a conclusion, according to Cicero, who established the concept of Antichthon in Roman literature. But it remained for the greatest geographer of antiquity, Claudius Ptolemaeus of Alexandria, or Ptolemy, to organize into a coherent geography what was known about the face of the earth.

In Ptolemy's time, about 150 A.D., the known world was a strip 70 degrees wide, lying mostly north of the Tropic of Cancer, with Cadiz on the west, India or Cathay on the east, savage lands of ice and snow to the north, and an impassable zone of fire to the south. Like Mela, Ptolemy held that, as a natural consequence of the earth's sphericity, India could be reached by sailing west from Spain.

These ideas of the spherical earth, of Antichthon, of reaching the east by sailing west were kept alive in the Middle Ages by monastic scholars. A fourth-century Roman monk, Macrobius, justified the existence of Antichthon by sheer reason, as had Cicero. Beatus of Liebana, a Spanish monk of the eighth century, drew a world map showing a southern continent. It was "desert land, neighboring and firm, but unknown to us."

The "Antichthones" themselves may have suspected the existence of a southern continent around the pole. About 650 A.D., according to the legends of Rarotonga, the Polynesian voyager Hui-te-rangiora, paddling southward in his canoe, *Te-ivi-o-atea,* saw "things like rocks whose summits pierced the skies" projecting above the frozen waste of "pia" (arrowroot). It may have been that an ancestor of the Maori people first encountered icebergs in the Antarctic Ocean and reached the ice pack.

Mercator's king-sized land

The persistence of the idea of a southern continent for two and one-half millennia before it was found is certainly one of the remarkable

Mappemonde in gores by Antonius Florianus shows the Mercator-Oronti

phenomena in the history of geography. With the concept of a spherical earth, this idea provided the intellectual basis for the expansion of Europeans into the western and southern hemispheres.

In the fourteenth century navigators believed that the temperate coasts of Africa south of the equator represented the northward extension of the fabled continent. The idea prevailed until Vasco da Gama sailed around the Cape of Good Hope in 1497 and found himself in the Indian Ocean.

As Europe awakened to the sciences, Ptolemy's geography was revived in the fifteenth century. A series of maps appeared in the cities of Europe, based on a treatise by Ptolemy which had been translated from Greek into Latin in 1409 by a Florentine scholar,

ncept of *Terra Australis* during the first half of the sixteenth century.

Jacobus Angelus. These were called "Ptolemaic" maps to indicate their origin in ancient authority. Some show the north and south poles. Antichthon appears in the southern part of the Indian Ocean. By the fifteenth century, however, mapmakers no longer called it by this ancient name. It had become Terra Australis, or Terra Australis Nondam Cognita—the Southern Land Not *Yet* Known.

The cartographic image of Terra Australis underwent an interesting evolution as reports of voyages into the new regions filtered back to geographers at home.

On his voyage of circumnavigation, Ferdinand Magellan saw the Land of Fires (Tierra del Fuego) to the south of his track as he beat through the stormy straits that bear his name at the tip of

South America. Geographers then assumed that Tierra del Fuego was nothing less than the northern tip of the legendary continent.

Early in the sixteenth century a definite conception of the size and contour of Terra Australis arose in Europe. It was strikingly modern. A continental outline similar to the one we know today, but with an exaggerated bulge in the Australian quadrant and a near connection with Patagonia, was first depicted by a French cartographer, Orontius (Oronce Fine), on a world map published in 1531. It next appeared on the famous *Orbis Imago,* or world map, of the Flemish mapmaker Gerhard Mercator (né Kremer) in 1538.

The southern continents of Orontius and Mercator in that period are identical. Both believed that Tierra del Fuego was the northern extremity of Terra Australis, which they showed as a somewhat circular land mass around the south pole, about one-third larger than its actual area. Considering the total absence of data, the Mercator-Orontius visualization of Antarctica in the first half of the sixteenth century is remarkable indeed. It was copied in Italy by Antonio Salamanca of Milan and Antonio Florian (Florianus) of Venice, who produced elaborate mappemondes based on the *Orbis Imago.*

But this amazing piece of guesswork did not last very long. In the second half of the sixteenth century the image of Terra Australis underwent a spurt in growth. Reports of new land sightings in the Pacific Ocean supported a much larger view of Terra Australis as a globe-girdling continent sprawling from the south pole to 45 degrees south latitude. This idea had been expressed in a map printed at Ulm in 1482 and on a globe made at Nuremberg in 1515 by Johannes Schoner, showing a huge southern continent called *Brasilie Regio.* Presumably influenced by the reports of new discoveries, Orontius and Mercator enlarged Terra Australis on their later maps, so that it spilled over the Tropic of Capricorn and extended at some points nearly to the equator.

By the end of the sixteenth century Terra Australis had grown to an area larger than that of the Americas. This king-sized continent is depicted in three great world maps: Mercator's 1587 map and the *Theatrum* and *Typus Orbis Terrarum* (1570 and 1587) of Abraham Ortelius of Antwerp. The image of a supercontinent persisted for more than two hundred years. It is found as late as 1730 in the *Ta'rikh al-Hind al-Gharbi,* a description of the New World printed at Constantinople.

The Spanish discoveries of the Solomon Islands, the Marianas, the New Hebrides, and the Marquesas seemed to confirm the idea that a temperate supercontinent, richer than America, was within reach. In England, Sir Richard Grenville petitioned Queen Elizabeth for a commission to explore "southward of the equinoctal [equator] to where the pole Antarctic hath any elevation above the horizon." At first Elizabeth vetoed the venture, as she did not want to antagonize Philip II of Spain, who claimed exclusive rights of exploration in the western hemisphere based on the division of new lands between Spain and Portugal by Pope Alexander VI in 1493. But after her relations with King Philip had deteriorated beyond repair, Elizabeth allowed Sir Francis Drake to pursue the project.

Drake's ship, the *Golden Hind,* was blown far south of Tierra del Fuego. There was no land in that wild and windy ocean. "The Atlantick Ocean and the South Sea meet in a most large and free scope," Drake reported. With this voyage the modern conception of Antarctica began to evolve. In one sense, Drake led the first Antarc-

Heart-shaped globe, engraved in Paris in 1575, depicts Terra Australis as an enormous land mass extending far north into temperate latitudes.

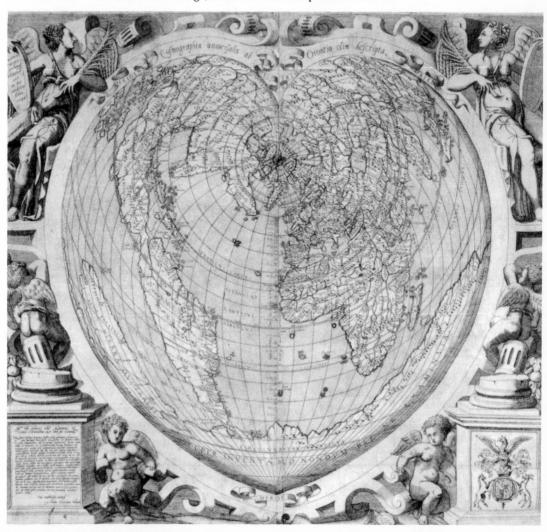

tic expedition, since it was the first voyage to have Terra Australis as its actual objective.[1]*

From that time on, the size of the supercontinent gradually diminished in men's minds as it eluded one expedition to the southern ocean after another. Yet the romantic idea of a temperate land in the south continued to intrigue colonizers, promoters, and even governments.

In 1766 Samuel Wallis sailed under sealed orders from the British Admiralty to seek new territory in the southern hemisphere and discovered for England the lovely island of Tahiti. The Chevalier de Bougainville reached it about the same time, and France promptly claimed it.

The famous voyages of Captain James Cook, the first European to sail across the Antarctic Circle, were made in the same spirit of seeing what there was to be had in the southern hemisphere grab bag. In 1768 he sailed for Tahiti, ostensibly to take British astronomers there to observe the transit of Venus across the sun in 1769. But he too carried secret orders from the Admiralty to search for Antarctica. If a southern continent did exist, there was no sign of it at 40 degrees south latitude. Alexander Dalrymple, a Scottish naturalist, urged the British government to keep trying. "The scraps from this table [in the south] would be sufficient to maintain the power, dominion, and sovereignty of Britain," said he. Why bother with the obstreperous colonies in North America, he demanded, when there obviously lay in the south a land more spacious than all of Asia?

Before long, however, this eighteenth-century view of Terra Australis became subject to a series of disillusionments. In 1773 a French navigator, Yves Joseph de Kerguelin-Tremarec, sought what he was sure was continental land containing wood, minerals, and even diamonds in the southern ocean. All he found was a rocky snow-covered island he called Terre Désolée, now known as Kerguelin Island, one of the larger sub-Antarctic islands in the Indian Ocean.

The great daydream was beginning to fade in the hard light of experience. On his second voyage, 1772–1775, Captain Cook crossed the Antarctic Circle three times. He reached the farthest south of anyone in the eighteenth century—latitude 71 degrees, 10 minutes, and longitude 106 degrees, 54 minutes west—300 miles north of the present Walgreen Coast. He saw nothing but the ragged ice pack, mists, and snow.

Had Cook made his southward probe farther to the east, he would

* Numbered reference notes begin on p. 289.

have discovered the Antarctic Peninsula. But for the remainder of the eighteenth century Antarctica was still *nondam cognita*.

The Panhandle pointing north

The intrepid Cook was stopped by the ice pack. It stretched ahead of his ship toward the south as far as anyone could see. "Since therefore we could not proceed one inch further south no other reason need be assigned for our tacking and stretching back to the north," he wrote. "No continent was to be found in this ocean but must be so far south as to be wholly inaccessible on account of ice."

In the autumn of 1774, Cook sighted the island of South Georgia. He claimed it for England—the first national claim made of an Antarctic territory. It was no land flowing with milk and honey. "The wild rocks raised their lofty summits until they were lost in the clouds," Cook observed. "And the valleys lay covered with everlasting snow. Not a tree was to be seen, nor a shrub big enough even to make a tooth pick."

The question of who first saw the Antarctic mainland has been in dispute for a century and a half. In Great Britain, Edward Bransfield and William Smith are credited with the first sighting of the continent about noon of January 29, 1820. It was not accidental.

Smith-Bransfield map of 1820, showing the routes of the brig Williams *in 1819 between Rio de la Plata and Valparaiso.*

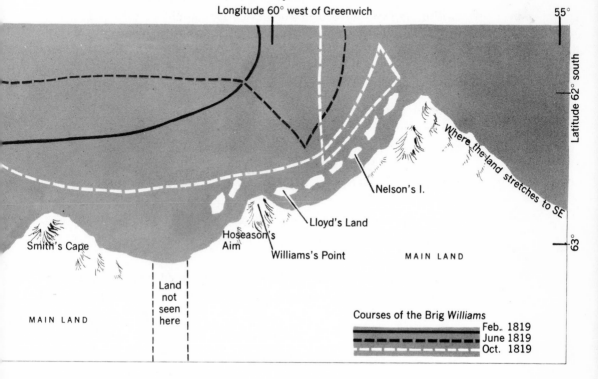

Longitude 60° west of Greenwich

55°

Latitude 62° south

Where the land stretches to SE

63°

Nelson's I.

Lloyd's Land

Hoseason's Aim

Williams's Point

MAIN LAND

Smith's Cape

MAIN LAND

Land not seen here

Courses of the Brig *Williams*
Feb. 1819
June 1819
Oct. 1819

While on a freight-hauling voyage around Cape Horn the previous year, Smith, in the brig *Williams,* had veered south to avoid a storm and had sighted snow-covered land (the South Shetland Islands). Ridiculed at first, his report later attracted the attention of the British Admiralty. Bransfield, a lieutenant in the Royal Navy, sailed from Valparaiso on December 20, 1819, in the brig *Andromache,* with Smith aboard.

On January 16 land appeared in the southeast, hard to distinguish from a layer of white clouds which floated above the summits of snowy mountains. Dr. Adam Young, the *Andromache*'s surgeon, described it as follows:

The whole line of coast appeared high, bold and rugged, rising abruptly from the sea in perpendicular, snowy cliffs except here and there where the naked face of a barren, black rock shewed itself amongst them. In the interior, the land or rather the snow sloped gradually and gently upwards into high hills which appeared to be situated some miles from the sea.[2]

The position of this sighting was given as 62 degrees, 26 minutes south latitude and 60 degrees, 54 minutes west longitude. In analyzing the voyage, Professor William H. Hobbs of the University of Michigan contends that Bransfield and Smith actually sighted one of the South Shetlands in this position.[3] Atlases published several years later in Europe and the United States indicated, however, that this sighting represented continental land.

If this had been the only sighting by Bransfield and Smith, the British claim to have seen Antarctica first might be invalidated, but it was not the only one. On January 29 they sighted land farther south, and their log indicates this was Trinity Island at the northernmost tip of the peninsula.

In the United States credit for first sighting of Antarctic continental land is generally given to Captain Nathaniel Brown Palmer of Stonington, Connecticut. At the age of twenty Palmer was master of the 44-ton sloop *Hero,* which accompanied a New England seal-hunting fleet to the region of the South Shetlands in 1820. Between November 15 and 21 he sailed the *Hero* south of the sealing grounds and sighted a stretch of the peninsula coast southeast of Deception Island on November 20, 1820. This sighting in latitude 63 degrees, 45 minutes south has been acclaimed by American scholars as the "first."

By a curious coincidence Palmer was credited with the discovery of the peninsula by a representative of the third party in the dispute,

Russia. The Soviet Union contends that Antarctica was discovered in 1820 by Captain Baron Fabian Gottlieb von Bellingshausen, who was searching for the southern continent on behalf of Czar Alexander I. According to one account,[4] Palmer's sloop encountered Von Bellingshausen's ships, the *Vostok* and the *Mirnyi,* shortly after sighting the peninsular coast. The Russian invited Palmer to come aboard the *Vostok* and listened attentively to the young New Englander's tale of the discovery. In retelling the incident, Von Bellingshausen referred to the coast as "Palmer's Land."

In 1907 Mrs. Richard Fanning Loper, Palmer's niece, published in the New London, Connecticut, *Globe,* a dramatized account of this meeting, quoting Palmer as follows:

I was ushered into the presence of the venerable commander, who was sitting at the table of his cabin, himself and a group of officers in full dress. The gray-headed mariner arose, took me by the hand, saying, through the medium of his interpreter, "You are welcome, young man, be seated."

I gave him an account of my voyage, tonnage of sloop, number of men and general details, when he said, "How far south have you been?" I gave him the latitude and longitude of my lowest point and told what I had discovered.

He rose much agitated, begging I would produce my logbook and chart, with which request I complied and a boat was sent for it. When the log book and chart were laid upon the table he examined them carefully without comment, then rose from his seat saying, "What do I see, and what do I hear from a boy in his teens—that he is commander of a tiny boat of the size of a launch of my frigate, has pushed his way to the pole through storm and ice and sought the point, I, in command of one of the best appointed fleets at the disposal of my august master, have for three, long, weary, anxious years searched day and night for."

With his hand on my head he added, "What shall I say to my master? What will he think of me?

"But be that as it may, my grief is your joy; wear your laurels with my sincere prayers for your welfare.

"I name the land you have discovered in honor of yourself, noble boy, Palmer's land."[5]

This account is not confirmed in Von Bellingshausen's report.

To Americans the mountainous land described by Palmer became known as the "Palmer Peninsula." On British maps, however, it was called "Graham Land," for Sir James R. G. Graham, First Lord of the Admiralty when it was discovered. Beginning 550 miles south of Tierra del Fuego, the peninsula curves southward to broaden out at its base into West Antarctica. The Chileans refer to it as O'Higgins

Land, named for the Chilean revolutionist Bernardo O'Higgins, who in 1818 liberated Chile. In 1964 an international convention agreed to call it the Antarctic Peninsula, but on a map it looks like a panhandle; hence I prefer to call it the Panhandle.

If Von Bellingshausen sighted the continental ice sheet, he made no such claim. However, he did claim for Russia a large island off the west coast of the Panhandle and named it Alexander I Island. Sailing around the continent, he visited South Georgia and the South Shetlands, where he observed that if the slaughter of fur seals continued at the rate of 60,000 to 80,000 a season they would not last much longer. He was right. They didn't.*

After two millennia, the southern continent had been glimpsed. It was no Elysian paradise. It was a fog-shrouded, terrifying region of ice and snow, empty, desolate, and silent. The motives for Antarctic exploration began to undergo an intellectual metamorphosis. The gold-and-diamonds motif faded. Prospects for colonization grew dim. In their place appeared other spurs, prestige and power. And a new force was stirring—science. Hitherto the Antarctic seas had been the preserve of the seal-butchers and the whale-hunters. Now the scientists began to infiltrate the region.

A fledgling United States of America took note of the widening world. In his first annual message to Congress, in 1825, President John Quincy Adams remarked that "the voyages of discovery . . . have not only redounded to . . . glory [of the nations that made them] but to the improvement of human knowledge. We have been partakers of that improvement and owe for it a sacred debt not only of gratitude but of equal . . . exertion in the common cause." But the young democracy was not yet ready to embark on a program of global exploration. "We have objects of useful investigation nearer home," Adams continued. "The interior of our own territories has been very imperfectly explored."

Private citizens, however, were interested in the Antarctic. Nathaniel Palmer joined two other men, an imaginative sea captain named Edmund Fanning and one Benjamin Pendleton, in promoting a voyage, raising funds principally from whaling and sealing interests. In 1830 the first American Antarctic expedition set sail in a ship called the *Annawan,* which cruised the seas to the north of the present Ellsworth Highland. Members of the expedition explored the South Shetland Islands, but did not sight the mainland.

* Von Bellingshausen's report was withheld for years. The Czar refused funds for its publication.

Aboard the *Annawan* was an Albany naturalist, James Eights, who was conducting research under a five-hundred-dollar grant from the New York Lyceum of Natural History (now the New York Academy of Science). He was the first to recognize as evidence of continental land to the south the displaced rocks and boulders, called erratics, that he found on the beaches of the sub-Antarctic islands visited by the *Annawan*. Such boulders were clearly continental in origin since they did not match the local outcrops. Where had they come from? Eights surmised they had been carried away from continental land in the south by icebergs, and then deposited on the islands by grounded bergs that had melted. If this was so, the bottom of the Antarctic seas must be liberally strewn with erratics. Eights's assumption was verified forty-four years later, when the dredge of the famous British oceanographic ship HMS *Challenger* lifted many of these rocks from the floor of the Antarctic Ocean.

In 1836, because of the feats of other nations in the Antarctic and the economic value of finding new sealing grounds, members of the United States Congress voted to finance an Antarctic exploring expedition. The story of the first official United States foray into Antarctic regions is not a brilliant one, but it was a beginning.

The expedition was nearly wrecked in its organizational stage by arguments about its composition, charges of graft against suppliers, and general mismanagement. On August 18, 1838, however, the expedition departed from Hampton Roads, Virginia, under the command of Lieutenant Charles Wilkes, but with only a handful of scientists. A number of outstanding American investigators who had signed up for the voyage withdrew when the Navy tactlessly advised them that their findings could not be published until they had received security clearance.

Wilkes's squadron consisted of six ships. None were fit for Antarctic waters, a fact the Navy could have ascertained if it had consulted the only Americans acquainted with the south polar seas—New England sealers and whalers. There were two 700-ton sloops, the *Vincennes* and the *Peacock*; the 200-ton gun brig *Porpoise*; and two tenders, one an antique New York pilot boat called the *Sea Gull,* and the other, called the *Flying Fish.* The storage ship *Relief* was so slow she had to be sent home before the squadron reached Antarctic waters.

Wilkes sighted the South Shetlands in March 1839 and sailed his ships along the eastern coast of the Panhandle. He attempted to push into the embayment now known as the Weddell Sea, but terrible clockwise currents and menacing ice drove him back. Water poured in through the open gun ports of his vessels. Clothing purchased by the Navy quartermaster at Philadelphia proved too thin for the intense cold. Wilkes was forced to retreat northward to Orange Harbor at Tierra del Fuego.

Curiously enough, the *Flying Fish,* dispatched on a probe southward, swept almost within sight of what is now called Marie Byrd Land—but she did not press quite far enough south—and returned with no report of any consequence.

The squadron reassembled at the South Shetlands and retreated to Valparaiso for repairs and some refitting. In May 1839 Wilkes sailed into the South Pacific and anchored his squadron in the harbor at Sydney, Australia. While the American sailors tried pathetically to make the ships weatherproof by rigging tarred canvas, word sped through the harbor that a magnificently equipped British expedition was coming to explore the Antarctic under the experienced arctic explorer Captain James Clark Ross. This did not help the already low American morale.

Wilkes sailed out of Sydney Harbor the day after Christmas, 1839, and headed south. On January 19, 1840, after a rapid voyage, he sighted a massive landscape to the south-southeast—unquestionably the coast of Oates Land, on the edge of East Antarctica. Driving the exhausted and sick crew on and on during February, he pushed his inadequate ships halfway around the perimeter of the coast of what is now called Wilkes Land. Beyond the ice cliffs he could see the rocks of low mountains, the brilliance of an ice sheet.

But his return to the United States was not greeted with great enthusiasm. Complaints about his treatment of his men and the conduct of the voyage had preceded him. Wilkes was promptly tried by court martial on charges growing out of harsh, peremptory disciplining of his men. He was convicted of illegally punishing subordinates and fined so heavily that he was nearly ruined financially.

But the indomitable Wilkes did not let this stop him. In 1844 he published privately his account of the expedition, offered it for sale, and recouped his financial losses when it became, overnight, a bestseller. The scientific results of the expedition appeared later in a small edition published by the government.

Some of the landfalls Wilkes reported were evidently erroneous, for they were never confirmed. But he was the first commander to identify the broad sweep of coastline the expedition sighted as "the Antarctic continent." Of course, this was a lucky guess. Wilkes had no way of knowing whether the land he saw through the fogs and mists was continental or insular.

Preoccupied with extending its own frontier, the United States made no further government-sponsored sorties in Antarctic waters in the nineteenth century. But American sealers and whalers were active everywhere in the southern seas.

During the mid-century period, the thinking about polar regions was influenced by the hollow-earth theories of John Cleves Symmes of Ohio. A retired infantry officer and onetime schoolteacher, Symmes published a pamphlet stating that "the Earth is hollow, habitable within, containing a number of concentrick spheres, one in the other, and it is open at the poles 12 or 16 degrees." He sent this declaration, with a certificate testifying to his sanity, to five hundred institutions of learning and government officials in the United States and Europe.

Symmes explained the polar openings as a result of the contraction of the earth's crust away from the poles, due to rotation. According to his theory, the entrance at the south pole to the center of the earth was 4000 miles wide, while at the north pole it was only 2000 miles.

The hollow-earth idea was not finally dismissed until Robert E. Peary reached the geographic north pole in 1909 and found nothing there but frozen sea ice.

Search for the south magnetic pole

In nineteenth-century Europe the Industrial Revolution was generating a new interest in science and technology. The expansion of maritime interests into polar seas challenged the art of shipbuilding, which had not changed basically in two millennia. Stouter ships, designed to breast the Antarctic ice pack and ride the storm-lashed southern ocean, began to appear, in response to the demands of the whaling industry. But better ship construction was not enough. In the second quarter of the nineteenth century national maritime ambitions required more efficient techniques of navigation.

Improved navigation, in turn, demanded a more precise knowledge of the earth's magnetic field and its inclination, or dip, toward the magnetic poles. The distinction between the magnetic poles and the geographic poles of the earth's spin axis was grasped by only a few astronomers, mathematicians, and master mariners. So far as navigation was concerned, it was more important to locate the magnetic poles than the geographic poles, so that the magnetic dip could be charted at all latitudes.

The concept of an invisible magnetic field over the earth had been held by seamen since the time of Columbus. The English scientist William Gilbert published in 1600 a hypothesis that magnetism was generated by the earth. Thereafter magnetic-field phenomena were studied intensively in Europe, and James Clark Ross had found the north magnetic pole in the Boothia Peninsula of the Canadian Arctic.

Once this was known, it was theoretically possible to locate the antipodal south magnetic pole. The German mathematician Johann Karl Friedrich Gauss predicted it would be found in latitude 66 degrees south and longitude 146 degrees east. If the south magnetic pole could be found where Gauss said it was, the practical application of magnetic-field theory to navigation would no longer rest on assumption.

In secrecy both Britain and France prepared to send expeditions to find the south magnetic pole. Why such a venture was classified is hard to imagine. King Louis Philippe of France commissioned Jules Sebastien César Dumont d'Urville to sail into the Weddell Sea and seek the magnetic pole. But d'Urville's attempt to enter those treacherous waters was no more successful than Wilkes's had been. It was just as well, for the magnetic pole was nowhere near there.

D'Urville then went to Polynesia, where he conducted an ethnological investigation. In the autumn of 1839 he turned south toward Antarctica to look for the magnetic pole near where Gauss had predicted it would be, and soon d'Urville and Wilkes were sailing the same seas, each unaware of the other.

On January 19, 1840, d'Urville sighted an icy coast he named Terre Adélie, after his wife—a doubtful tribute. The sighting was made the same day as Wilkes's first landfall, and both expeditions saw the same coast. Later the Frenchman's ships, the *Astrolabe* and the *Zélée,* encountered the *Porpoise* of Wilkes's squadron in a dense fog at 65 degrees south latitude and 135 degrees east. The ships sailed by each other, flags flying, crews watchful but silent, until each vanished from the sight of the other in the swirling mist. Neither expedition actually found the magnetic pole, since it was located on land. In the same year the British expedition led by Captain Ross was also seeking the same pole. Ross arrived in Tasmania in two reinforced ships, the *Erebus* and the *Terror,* with double decks and double hulls, sheathed in copper, braced inside with squared beams, containing watertight compartments. It was the best-equipped polar expedition yet dispatched to seek the south magnetic pole and the southern continent.

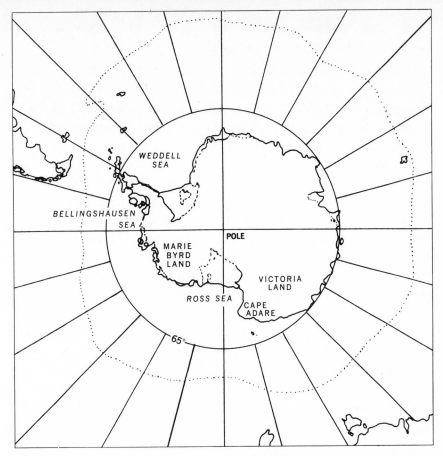

As spring came to the southern hemisphere in 1840, Ross's ships plunged southward into the wild Antarctic Ocean along the meridian of 170 degrees east. It was a lucky guess, this tack, for Ross entered the more accessible of the two great Antarctic embayments, now known as the Ross Sea. Once his stout ships had broken through the ice pack and emerged into the blue-black deeps of open water, Ross had relatively clear sailing to the south for nearly 400 miles.

What a sail it was! To the west reared up stupendous peaks, some covered with snow that flashed in the sunlight, others bald and scoured by a long-gone glacier, and a few knifing into the sky like weapons. Ross's ships passed Cape Adare (as it is now called) and entered into the heartland of Antarctica.

The great mountain system to the west guarded awesome evidence of vast lands beyond it. Massive glaciers, rivers of ice, flowed in slow motion between the peaks, down to the sea. The weather was clear and fine, as dry and bracingly cold as Antarctic weather can be. The sun shone all day and all night, illuminating a landscape entirely white and blue, majestic with a grandeur and a beauty no other land can equal. By comparison, Ross observed, the Arctic seemed dismal indeed.

At Cape Adare, Ross took readings. According to the dip of his compass, the magnetic pole lay in latitude 76 degrees south and longitude 145 degrees and 20 minutes east—a distance of about 500 miles southwest of the cape. Could he sail there? Obviously, he could not, at this latitude. But somewhere there must be a water passage.

As the British geographer and polar historian L. P. Kirwan has pointed out,[6] Ross was thinking of the region in terms of his experience in the Arctic. There one sailed around and between great islands. In spite of the ancient hypothesis of a continent, Ross thought of Antarctica as analogous to the Arctic. The mountains and the regions lying beyond them he had named Victoria Land after his queen. But he thought of them not as continental land but rather as a group of large, mountainous islands, through which, at any moment, lookouts might spy a water passage. After all, so much of the land that other mariners had called continental had turned out to be insular that it would have been no surprise to anyone if nothing but islands existed in the southern ocean.

By pressing south, Ross sought to sail around the mountains. It was fine weather for sailing, as the austral summer of 1840–1841 was exceptionally mild, with little precipitation. By Ross's accounts, the sea was wide and open.

Day after day the *Erebus* and the *Terror* plowed through this splendid sea. To the east a large island formed on the horizon, crowned by one of the strangest sights in the world—a great white cone rearing skyward from the island. From the top of the cone, which the mariners could see had been truncated, rose soft billows of white smoke. A snow-covered active volcano, surrounded by ice, was an image worthy of all the descriptive power of Coleridge, who thought of snow and ice as some woeful, wintry hell. Yet here was a sight, this great volcano of the Antarctic, that exceeded the Coleridgean imaginings. It was not fearful but majestic and marvelous. Beneath the icy mantle burned the fires of the inner earth.

Ross named the volcano Mount Erebus. As the wooden sailing ships moved near, the volcano became alive, belching heavy smoke and flame. Beyond it, white and inert, rose another great cone, which he named Mount Terror.

To the west of the island, now known as Ross Island, his ships moved into a wide sound, now called McMurdo Sound. The western

Mount Erebus on Ross Island, partially obscured by clouds, in a phot graph taken in 1911 by Herbert G. Ponting of the Terra Nova *expeditio*

shore of the sound remained the unbroken range of mountains that the ships had followed for four hundred miles. But Ross and his associate commander, Captain Francis M. Crozier in the *Terror*, expected to reach open sea south of the island.

Instead, they sighted another fantastic and unexpected sight. Across the sea ahead of them a massive wall of ice took shape. They sailed up to it and hove to. The barrier rose 150 feet. As far as the eye could see it stretched across the sound from the island to the mountains.

Ross and Crozier sailed their ships around the island, backtracking to the north, for there was no passage southward. They then pushed eastward along the great barrier of ice, nosing for a passage, like terriers confronted by a fence too high to jump. There was no way through.

Before them lay the largest ice shelf in the world, now known as the Ross Ice Shelf, a sheet of floating ice the size of France, with Belgium and Luxembourg thrown in. To Ross and his men it was the "Great Barrier," later to be called the "Ross Barrier." More than a century was to pass before the nature of this greatest of all floating ice sheets on earth was to yield to man's investigation.

Now there was no way south or west. The magnetic pole lay still far to the west, beyond the gleaming mountains, somewhere in the ice fields that now and then could be glimpsed at the head of a glacier.

Ross wisely sailed northward as the austral summer came to an end in March 1841. His ships reached Hobart, Tasmania, on April 6, after 145 days in the Antarctic.

On this one expedition, Ross had discovered the Antarctic as no man before him had. Wilkes and d'Urville had sighted the coast, but from a distance. The *Erebus* and the *Terror* penetrated the region to the Great Barrier. Ross had found the way in. Yet it was not until sixty years later that one of his countrymen, Robert Falcon Scott, was to follow up this penetration of the new land with the first extensive explorations on the grounded ice cap itself.

Landfall

For several decades there were no major Antarctic voyages for the sake of exploration. Scandinavian, British, and American seal-hunters continued the slaughter of the fur seals until by 1870 the species had become virtually extinct.

The next surge of exploration came at the end of the nineteenth century and paved the way for opening up the region to man in the twentieth. The motives were both economic and scientific. The Greenland whaling industry was faltering. The right whale was disappearing. British and Norwegian whalers turned to the Antarctic, where the great whales fed on the rich plankton of the south polar seas, and scientists attached themselves as supercargo to the whaling sorties in the southern ocean.

In 1893 Captain C. A. Larsen, a Norwegian sailor with a scientist's eye for observation, sailed to the South Orkney Islands, and from there into the Weddell Sea. It was Larsen who found on an island off the northern Panhandle some curiously carved balls of stone which bespoke human occupation, perhaps only brief, of this region at some time in the past, possibly by Patagonians.

Another Norwegian skipper, Leonard Kristensen, sailed into the Ross Sea. No such exotic evidence of human visitation was found there, however. If men had once migrated southward down the Panhandle, it is not likely they reached the main sheet of continental ice, nor is it reasonable to suspect they would have remained long if they had. The mystery of the stone spheres has never been solved. Someone had passed through the Panhandle. But who? And when?

Aboard Kristensen's ship, the *Antarctic,* were a Norwegian explorer, H. J. Bull, and an Australian schoolteacher, Carsten E. Borchgrevink. They led a landing party ashore on the stony beach of Cape Adare, in the shadows of the great mountains, on January 24, 1895. The date is a memorable one because this was the first time man had set foot on the seventh continent.

After that came a series of "firsts." The first party to winter over in an Antarctic region did so involuntarily. It was the Belgian expedition of Adrien de Gerlache. His ship, the *Belgica,* was beset in the ice off the coast of Alexander I Island during the austral winter of 1898. Aboard the *Belgica* were two men whose names were to resound in the polar halls of fame. The first mate was Roald Amundsen, and the surgeon was Dr. Frederick Cook, American physician and anthropologist, whose claim in later years to having been first at the north pole was to set off a controversy which has not yet entirely died down.

In 1899, Borchgrevink, with the financial support of an English magazine publisher, Sir George Newnes, sailed into the Ross Sea in the *Southern Cross.* Borchgrevink headed the first party of men to winter over on continental Antarctica. The quarters consisted of a rude shack roofed with tarred canvas, but all survived the Antarctic winter. And these first "Antarcticans" painstakingly assembled the first continuous record of magnetic, weather, and auroral observations during the austral winter.

When spring came, Borchgrevink and several members of the party made their way southward to McMurdo Sound. They climbed up the great Ross Barrier and stared into the seemingly infinite distance on the apparently endless plain of ice that stretched into the horizon to the south and east.

They made the first sled trip over the ice shelf to latitude 78 degrees and 50 minutes south. It was a new "farthest south" record for mankind. It was also a long step toward the geographic south pole.

The year 1900 was an active one in the Antarctic. A German expedition, led by Professor Erich von Drygalski, explored the Atlantic coast of the continent. A Swedish geologist, Otto Nordenskjold, led an expedition to the Antarctic Panhandle. After two years of hard-bitten investigation, Nordenskjold concluded that, geologically, the Panhandle was a part of South America—not of Antarctica. Most geologists today agree with him. He was the first to formulate this conclusion and pronounce it before the world. Was Antarctica one land mass, or two?

The Edinburgh naturalist William S. Bruce led a Scottish National Antarctic expedition into the hostile Weddell Sea. He sighted the white emptiness of Coats Land, to the east of the ice shelf, which fills the southern half of the Weddell Sea, just as the Ross Barrier fills the southern half of the Ross Sea.

In England, meanwhile, a conviction was growing in some echelons of government that the time had come to plunge boldly into the Antarctic with a large, well-stocked, and ably led expedition. Thus far, since 1820, only the fringes of the new continent had been penetrated. Men had barely glimpsed it; then they had set foot on it, only to pull back. The time was ripe to penetrate it. For who knew what lay beyond the glistening mountains or the great ice barrier?

Murray's hypothetical continent

When the twentieth century dawned, the nature of Antarctica was still in shadow. It was the last major physiographic problem on the earth. Perhaps the extent to which this was so has been appreciated only in recent years, which have shown the magnitude of the mystery.

The classical and medieval idea of a southern continent had persisted. Here to all appearances was the Antarctic continent, virginal in its shield of ice. But the experiences of seafarers still suggested that in the ice and mists of the south lay only large islands between great frozen wastes of ocean. Whalers and sealers had hunted in Antarctic waters for more than a century, but all the landfalls they had made were islands. There were, to be sure, the unfathomable coasts perceived by Wilkes and the shining mountains of Victoria Land which Ross had admired. Were these merely larger islands in a frozen sea?

Yet proof that a lost continent did indeed lie beneath the ice had been found on the first American scientific journey into the Antarctic in 1830 by James Eights in the *Annawan*.

The evidence for a continental Antarctica was summarized on the evening of February 24, 1898, in the gaslit hall of the Royal Society of London. The speaker was Dr. John Murray, a Canadian oceanographer,[7] who had served as a biologist on the world-wide scientific cruise of the British research ship *Challenger* from 1872 to 1875. The steam-driven corvette had crossed the Antarctic Circle in 1874 and skirted the edge of the ice pack. While no one aboard saw land or the icebound coast, Murray described huge icebergs floating northward in the cold gray Pacific Ocean. He said:

These flat-topped icebergs form the most striking peculiarity of the Antarctic Ocean. Their form and structure seem clearly to indicate that they were formed over an extended land surface and have been pushed over low-lying coasts into the sea.

As these bergs are floated to the north and broken up in warmer latitudes, they distribute over the floor of the ocean a large quantity of glaciated rock fragments and land detritus. These materials were dredged up by the *Challenger* in considerable quantity and they show that the rocks over which the Antarctic land ice moved were gneisses, granites, micaschists, quartziferous diorites, sandstones, limestones, and shales.

These lithological types are distinctively indicative of continental land and there can be no doubt of their having been transported from land situated near the South Pole.

The elegance of this reasoning made a strong impression on the assembly. Most of the scientists there were disposed by a classical education to a belief in the actuality of a southern continent.

Murray cited similar evidence found by others in the southern sea. Had not Dumont d'Urville in 1838 described the rocky islets off the coast of Adélie Land as composed of continental granite and gneiss? Had not Wilkes found an iceberg near Adélie Land containing

boulders of red sandstone and basalt? Borchgrevink had brought back fragments of mica-schist and other continental rocks from Cape Adare in 1895. Wasn't this further evidence?

The identification of granitic and schist rocks from the misty regions of the south pole was as exciting as such a finding might be today on the planet Mars. This was the same kind of rock basement into which the engineers of the 1890s were drilling to anchor the skyscrapers of New York and Chicago.

"We are thus in possession of abundant indications that there is a wide extent of continental land within the icebound regions of the southern hemisphere," Murray concluded.

So there arose "Murray's hypothetical continent," which gave scientific support for the ancient belief in the continentality of Antarctica and was a powerful stimulus to further exploration by England. It provided a challenge to burgeoning British science, as well as to an ancient spirit of derring-do in that land, in the closing years of the reign of Queen Victoria. In the first decade of the twentieth century a series of British national expeditions opened up the heart of Antarctica and the way to the south pole.

"These great southern solitudes"

The first deep penetration of the seventh continent, as Antarctica was coming to be known early in the twentieth century, was made by the British *Discovery* expedition. Its commander, Robert Falcon Scott, a lieutenant in the Royal Navy, was only 33 years old when the destiny of 123 men, a large ship, and an important mission were entrusted to him. He had never before visited a polar region or sailed a ship in a polar sea.

Nevertheless, Scott proved from the outset that he was an energetic and capable leader. His mission was clearly spelled out—to determine the nature and extent of the Antarctic continent and the thickness of the ice that covered it, though these questions were to elude science and exploration for another half century.

Considering the vastness of the project, which was underestimated in the England of 1900, Scott made a creditable beginning. The end, however, has not yet been reached and it is doubtful that it will be in this century. The last land frontier on earth has proved to be, perhaps, the most difficult and intricate to fathom.

The shipyards at Dundee, Scotland, turned out a ship for the expedition, the *Discovery,* after which the expedition was named. Scott

and his complement sailed from the London docks on July 31, 1901. At Cowes, downstream, King Edward VII and Queen Alexandra visited the ship. Scott never forgot this gracious act. He wrote of it even as he lay dying eleven years later in a snow-buried tent on the Ross Ice Shelf.

Knowing no better way to proceed to Antarctica, Scott followed the tracks of Cook and Ross into the southern ocean. By November 15 the *Discovery* had passed the 60th parallel south. On the morning of November 16 the first sea ice appeared. Scott had never seen such a sight before. The words he committed to his diary to record his impressions give us some insight into the poetic nature of this determined and introspective man, who was one of the greatest of all explorers.

"The wind had died away," he wrote. "What light remained was reflected in a ghostly glimmer from the white surface of the pack. For the first time, we felt something of the solemnity of these great southern solitudes."

The *Discovery* swung across the Pacific Ocean to the northwest to make port at Lyttleton Heads, the harbor for the city of Christchurch on the South Island of New Zealand. Port Lyttleton is still the jumping off point for Antarctica-bound vessels.

Scott ordered supplies for a two-year journey, including pre-cut lumber for the huts he planned to erect on land as wintering-over shelters. Then he sailed southward in December on the 1800-mile sea route to the ice cap.

Land was sighted at 10:30 p.m. on January 8, 1902. It was the northern rim of the Trans-Antarctic Mountain Range, which guards the ice fields of Victoria Land. Scott and his men stared at one of the most awesome sights on this planet, as the cold wind off the ice cap dashed icy spray into their faces and newly grown beards. Scott wrote of the night scene, "The sun near the southern horizon still shone in a cloudless sky and far away to the southwest the blue outline of the high peaks of Victoria Land could be seen."

He turned the *Discovery* inshore at stony Cape Adare, not far from the site of Borchgrevink's camp. Piles of loose rocks called moraines on the spit of barren beach gave testimony "to vaster ice in former ages," he observed. This was one of the earliest suggestions that, at least in the Ross Sea region, the ice has retreated.

Sailing southward once more, the *Discovery* approached the great Ross Barrier, and turned to the east. The ship dropped anchor off Ross Island. With Dr. Edward A. Wilson, the expedition's physician and zoologist, and Lieutenant Charles W. R. Royds, Scott climbed a volcanic cone at Cape Crozier, one of the numerous extrusions of land from Ross Island. They peered from the top of the high cone out over the ice barrier, expecting it to be a flat area of ice, perhaps giving way to other islands in the south and east. But the surface of the great ice shelf was too undulating to permit a view of more than a few miles. There was no way of judging the size of the ice shelf by visual inspection—at least not from Cape Crozier.

During the last week of January the *Discovery,* with sails furled, steamed slowly eastward along the barrier ice. There was no sign of the "Parry Mountains" which Ross had believed he had seen.

At the end of January the expedition reached a portion of the barrier which dropped in elevation to within 10 and 20 feet of the sea. Beyond it they saw snow slopes rising to the east. Soundings from the ship showed shallow water. Here was land at the eastern edge of the barrier. Scott named it King Edward VII Land.

On February 4, 1902, Scott climbed into the basket of a gas balloon and ascended to an altitude of about 1000 feet—the first aerial reconnaissance in Antarctica. What he saw was the endless, undulating plain of the ice shelf, not flat as he and others had expected, but a sequence of long, wavy slopes.

Now the summer season in the southern continent was far advanced, and young ice was forming at the edges of the barrier. Scott sailed back west to McMurdo Sound. Winter quarters were erected of the pre-cut lumber at Cape Armitage, another volcanic extrusion of the island into the sea. To the west, across forty to fifty miles of the ice-dotted waters of the sound, rose the mountains of Victoria Land, while to the east of the camp towered the white-mantled, smoking volcano Mount Erebus. Never before had men prepared to dig in for the occupation of such hostile, yet scenic surroundings.

The *Discovery* hut, built on a portion of the cape named "Hut Point," provided housing for men and shelter for dogs and ponies. They would set out, when spring came the following October, to explore the new land. Soon the 24-hour sun began to set on a diurnal cycle. Sub-zero gales of late fall swirled snow into the encampment. The *Discovery* became fast frozen in the ice of the sound. The dark night of winter descended on this lonely outpost of humanity in Antarctica.

When the sun returned, on August 21, 1902, Scott climbed to the top of Crater Hill nearby to see the spectacle of an Antarctic sunrise. "Over all the magnificent view," he wrote, "the sunlight spreads with gorgeous effect after its long absence, a soft pink envelops the western ranges, a brilliant red gold covers the northern sky. To the north east crystals of snow sparkle with reflected light."

The austral spring came, and the air warmed. During November and December, Scott led a sled party southward over the shelf to 82 degrees, 16 minutes south, a round trip of about 960 miles from Hut Point. The distance was covered in 93 days. Serious attacks of scurvy broke out. Had the party not returned to base on time, the attacks would have weakened the men fatally. One of them, Ernest Henry Shackleton, became so ill he had to be invalided home in the relief ship. He later became Scott's rival in Antarctic exploration.

A view of Hut Point on McMurdo Sound, where Scott built his 1902 winter encampment, with the cone of Mount Erebus in the background.

Late in December, the relief ship, the *Morning,* commanded by Lieutenant William Colbeck, arrived. The *Discovery* remained a prisoner of the ice that summer. On March 2, 1902, the *Morning* departed for England, and Scott and his men settled down for their second winter on the ice.

Scott was determined to climb one of the mountain glaciers through the high escarpment of Victoria Land, to take a look at that region. While his sled party had been pushing southward on the ice shelf earlier in the year, Lieutenant Albert B. Armitage and Reginald W. Skelton, the *Discovery*'s chief engineer, had crossed the sound and had found a way up to the high plateau in latitude 78 degrees south via a glacier which ascended some 9000 feet—now called the Skelton Glacier.

A look at Victoria Land, then, became the main object of Scott's second major journey. By October 12, 1903, the weather had moderated to permit Scott and twelve men to set out from Hut Point. Advance scouting indicated that the Ferrar Glacier,* a river of ice that flows through the mountains to the south of the Kukri Hills, might be more accessible to the expedition's eleven-foot sleds than the glacier which Armitage and Skelton had climbed.

It was 90 miles up the Ferrar Glacier on ice as hard as steel. At the end of 6 days, the party had climbed to an elevation of 6000 feet. The German silver they had used to coat the runners of the sleds had worn off all but one. Since the runners would not hold up on the hard ice without the coating, the entire party was forced to return to Hut Point for repairs.

A week later they were off again. By November 1 they had reached the food cache they had installed on the first sortie, near the top of the glacier. There a minor tragedy befell them. A high wind sweeping off the plateau blew away Scott's *Hints to Travelers,* a useful book containing tables which enabled mariners to find latitude and longitude by sun and star sightings anywhere on earth.

When they at last reached the head of the glacier, the men were pinned down for seven days by a ferocious blowing snowstorm. They did not arrive at the high plateau country of Victoria Land—9000 feet above sea level—until mid-November. The plateau was incredibly cold, though it was late spring. In the face of continuing high winds and temperatures ranging down to 40 degrees below zero Fahrenheit, the party turned back.

* Named for Hartley T. Ferrar, the expedition's geologist.

On the return journey Scott and Lieutenant Edgar Evans, at the head of the column, saw their sled and dogs suddenly vanish in a puff of powdery snow. Before they could brace themselves, the two men were sliding downward after the sled and the dogs into the blue grotto of a great crevasse.

Scott and Evans saved themselves from falling to the bottom of the icy chasm by clinging to the traces, which held the dogs suspended in air, as the sled had wedged itself between the narrow walls of the crevasse. "There were blue walls on either side and a very horrid-looking gulf below," said Scott. Slowly and painfully men and dogs were extricated by the others. The expedition returned to Hut Point on Christmas Eve of 1902, after a trek of 59 days and 725 miles.

Meanwhile another party, led by Lieutenants Michael Barne and George F. A. Mulock, had followed the Victoria Land coastline southward along the ice shelf. It was becoming apparent to the explorers that the shelf floated upon the southern portion of the Ross Sea. But the true nature of the shelf as an extrusion of the grounded ice cap of West Antarctica was not to become established for more than fifty years. Barne and Mulock fixed the positions of about two hundred mountain peaks in this snowy range. Their observations showed that the shelf was moving slowly northward.

Early in January of 1904, two relief ships appeared—the *Morning* and the *Terra Nova*. They anchored at the edge of the ice in McMurdo Sound and delivered to Scott his orders from the Admiralty to return to England, with the *Discovery,* if she could be pried out of the ice which had held her fast for two years, or without her if not.

That year twenty miles of solid ice lay between the *Discovery* and open water. Reluctantly, the commander and his men began to transfer supplies to the relief ships. Other teams chopped and dynamited the ice. Toward the middle of February great cracks suddenly appeared in the ice, and with explosions that sounded like bombs it began breaking up quite rapidly, until at last the *Discovery* once more floated on open water.

Before leaving Ross Island, Scott's party erected a wooden cross on a ridge at Hut Point in memory of a seaman named George T. Vince, who had been lost in a blowing snowstorm the first year. The cross is still there—the first memorial to death in the Antarctic. "In this undecaying climate," Scott remarked in his account of the expedition, "it will stand for centuries."

Sailing northward to New Zealand, Scott searched for a portion of

the coastline Wilkes had charted in 1840, but there was no sign of it. Presumably Wilkes had mistaken low-lying fog banks for land.

On April 1, 1904, the *Discovery* sailed into the harbor at Lyttleton. Most of the population of Christchurch, then a city of more than one hundred thousand, turned out to acclaim the return of the heroes. The southern continent had been claimed by man. The last of the planet's land frontiers was open. Antarctica could no longer be called *terra nondam cognita*.

Strange formations seen at the junction of Ross Island and the Ice Shelf.

Overleaf:
A penguin rookery at Cape Royds on McMurdo Sound, photographed by Herbert G. Ponting, member of the Scott Terra Nova *expedition (1910–13).*

THE SEVENTH

CONTINENT

The *Discovery* expedition presented to the world the first generalized picture of Antarctica, a blue-white glacial land immersed in ancient, but receding, ice. Scott's principal findings were geographic rather than scientific.

At the eastern edge of the Ross Ice Shelf the expedition had discovered King Edward VII Land, or what seemed to be land. It had made a large-scale reconnaissance of the shelf. It had explored the McMurdo Sound region and had traced the course of the Trans-Antarctic Mountains for nearly eight hundred miles. It had also made the first penetration of the high plateau of Victoria Land.

Scott himself was greatly impressed by the evidence he saw of ice retreat, from Cape Hallett, where a series of moraines showed that the ice had been more extensive in time past, to McMurdo Sound. On the mainland coast of the sound there were several dry valleys from which the ice had retreated unknown ages ago. The ice had left its mark high on valley walls; now it was gone. From these observations and others, a clear idea that there had been ice recession in Antarctica was generally formed, although no such claim was made officially. This, however, does appear to have been a prevailing opinion, and it made a profound impression upon students of geophysics and climatology.

The fact that the *Discovery* expedition was operating within the boundaries, for the most part, of the largest ice-free region in Antarctica was not then realized, nor were the special conditions which influence local climate in the McMurdo Sound region taken into account.

It remained for another British national expedition to take a step farther south and probe the polar plateau. The leader was Ernest H. Shackleton, who had suffered the humiliation of having been invalided home in 1903 aboard the relief ship *Morning*. Along with Scott and Dr. Wilson, Shackleton had suffered severely from scurvy on the ice-shelf traverse. Unlike his companions, the burly and robust Shackleton had not recovered satisfactorily after the party staggered back to Hut Point.

So Shackleton had gone home. Scott understood the man's feelings and sympathized with him. The experience seemed to leave Shackleton with a bitter compulsion to master the hardships and frustrations of this region. He kept coming back in future years until at last, twenty years later, he died in the attempt.

In 1907 Shackleton raised funds for the second British foray into the seventh continent. Early in January 1908 he left New Zealand in a small whaling ship, the *Nimrod,* which was towed as far as the Antarctic Circle to save coal.

Shackleton established his base at Cape Royds, another volcanic extrusion of Ross Island into McMurdo Sound about eighteen miles from Hut Point. The volcanic beach was inhabited by hundreds of Adélie penguins, who stood their ground when Shackleton's men

A group of Adélie penguins and chicks at the Cape Royds rookery.

hauled supplies ashore. The little Adélies are still there. So is Shackleton's hut, with much of the canned food and miscellaneous pieces of equipment his party left there, still intact after half a century. The *Discovery* hut at Hut Point still stands too, but a good deal of restoration work has been done on it.

Shackleton was the first expedition leader to try Manchurian ponies on the ice cap. Any competent veterinarian should have been able to tell him their hoofs were not designed to travel over glacier ice. This became apparent when he and his men attempted to drive pony-drawn sledges up the 100-mile-long Beardmore Glacier to the 9000-foot-high south polar plateau. One by one the exhausted ponies sank down to the ice and had to be shot. Before the party turned back, 110 miles from the south pole, the last of the ponies had been dispatched.

Shackleton's polar sortie established a new "farthest south" record, to 88 degrees, 23 minutes south latitude. Had he used dogs, he might have succeeded in reaching 90 degrees south—the geographic pole.

The expedition achieved other notable results. A party led by the Australian scientist T. W. Edgeworth David climbed Mount Erebus and looked down into the crater of that fantastic volcano, which still smokes under its icy mantle. During the second year of the expedition, in 1909–1910, David led a sled party to the south magnetic pole, which was then in 72 degrees, 25 minutes south latitude and 155 degrees, 16 minutes east longitude—a journey of nearly 700 miles from Ross Island.

The south magnetic pole continually moves around, with changes in the earth's magnetic field. By 1960, for example, the magnetic pole had drifted more than 1000 miles from where David found it, to latitude 67 south and longitude 147 east, near Cape Derision, Commonwealth Bay, on the Adélie coast.[1]

The French too were active in Antarctic explorations during the first decade of the twentieth century. An expedition in 1904–1905, under the scientist-yachtsman Jean B. Charcot, defined the coastline of the American quadrant of Antarctica (opposite South America) from the base of the Panhandle to King Edward VII Land. Charcot completed and refined these measurements on a second expedition in 1908–1910. He commanded a motor ship called the *Pourquoi Pas?* (Why Not?). The name seemed to express a whimsical attitude toward polar exploration, compounded of curiosity and scholarship, in contrast to the compulsive, morbid concern of Shackleton with the mastery of the region.

If the French were more relaxed about Antarctic exploration, it did not show in their equipment. The *Pourquoi Pas?* provided individual staterooms and workrooms for eight scientists who went along in addition to her crew of twenty-two. She had electric lights. It is said she stocked the finest wines ever shipped into polar regions. Charcot's expeditions were financed chiefly by the government, through the intervention of Aristide Briand, then Minister of Public Instruction. Briand viewed the Antarctic with the statesman's concern over its eventual disposition, as well as with the scholar's interest in its intellectual challenge.

In later years Charcot turned to Arctic regions. In July 1936 the *Pourquoi Pas?* was caught in a gale off the coast of Iceland. Dismasted by the fury of winds and seas, she was hurled on the rocks near Reykjavik. Charcot was last seen on the bridge of his famous ship, wearing a life jacket and directing the evacuation.

The polar race

In the power structure of the West in the early 1900s, the values driving men to the poles of the earth were not too far from those which impel us to reach the moon today. The north and south poles were the prestige goals of half a century ago. After the fifty-two-year-old Robert E. Peary, his Negro companion Matthew Henson, and four Eskimos reached the north pole on April 6, 1909, there remained only the south pole to test the spirit of man and the mettle of nations. Like space exploration today, polar exploration in the first decade of this century reflected material as well as scientific values. It depended on technology as well as on human endurance and skill. Its success measured the prowess not only of one group of men, but of a whole people. We can hardly say that our motives in Project Apollo, the program to place astronauts on the moon and bring them home safely, are basically different.

The mélange of motives and influences, both national and personal, which characterized these Antarctic expeditions, and probably all others, may have their basis in the biological drive of human beings to expand and control the physical environment. Antarctica was the last of the land frontiers on the planet. No one seriously believed by then that any habitable lands or precious metals would be found in available quantities there. No one seriously thought of the cold, beautiful icescape as an Eldorado. Nevertheless, in Great Britain, in Germany, and in Scandinavia, the restless drive to probe this region was stirring more feverishly than ever. Now the glorious goal was the south pole.

In 1910, the popular motive for reaching the south pole was analogous to the incentive in 1960 for reaching the moon. Scientifically, just getting there had no meaning. Only prestige was involved. Yet the race for this imaginary point on the ice cap attracted financing and popular support for another British national expedition. The pole, in fact, had acquired something of the emotional aura of the Holy Grail when Scott set sail on his second expedition to Antarctica from Cardiff, Wales, on June 15, 1910.

The expedition ship was the *Terra Nova,* built for whaling in the polar seas by the master shipwrights of Dundee. Scott knew the *Terra Nova,* which had been one of the relief ships sent to fetch him home in 1904, when it looked as though the *Discovery* would never be free of her icy prison. In recognition of the new technology, the

The expedition's photographer Herbert G. Ponting caught the Terra Nova *heeling in Antarctic waters on its voyage southward to McMurdo Sound.*

expedition carried three motor sleds which Scott proposed to try out on the ice.

Meanwhile, the Norwegians were not idle. Roald Amundsen, the Norwegian explorer who had served as first mate on the *Belgica* in 1898, had acquired the *Fram,* a famous Arctic polar vessel, from his guide and mentor, Fridtjof Nansen. Amundsen announced he was sailing to the Arctic via the Bering Strait. In the summer of 1910 the *Fram* was seen riding at anchor in the harbor of Buenos Aires, supposedly headed for the Strait of Magellan and the Pacific Ocean route to the northland.

When the *Terra Nova* reached Melbourne, Australia, there was a telegram awaiting Captain Scott. "Beg leave to inform you proceeding Antarctica. Amundsen," it said.[2] That was the first inkling Scott had that he was in a race for the south pole.

Early in January 1911 the *Terra Nova*'s lookouts sighted the smoky plume of Mount Erebus on Ross Island. A party put ashore to look into the condition of the old *Discovery* expedition hut at Hut Point. Shackleton's men had used it in 1907, but now Scott's party found it in poor condition. Scott moved the ship eighteen miles north to another volcanic spur called Cape Evans. Here, in the shadow of Erebus, the men of the *Terra Nova* built a new hut.

The weather was bright and sunny during January. Scott wrote that he chose the Cape Evans site "so that always towering above us we have the grand, snowy peak, with its smoking summit," and described the setting as follows.

The sea is blue before us, dotted with shining bergs or ice floes, whilst far over the [McMurdo] Sound, yet so bold and magnificent as to appear near, stand the beautiful western mountains with their numerous lofty peaks, their deep glacial valleys and clear-cut scarps, a vision of mountain scenery that can have few rivals.

During the preparation of the base a biological discovery was made which since has been confirmed many times. While standing on an ice floe in the sound, the expedition's photographer, Herbert G. Ponting, and several others were attacked by killer whales. The men were amazed at the efficient manner in which six of the carnivorous cetaceans worked as a team to break up and upset the floe, so that the men and sled dogs would tumble into the water. "It is clear that these creatures are endowed with singular intelligence and in the future we shall treat that intelligence with every respect," wrote Ponting. Modern research has established that the cetaceans (whales, dolphins) represent the intelligentsia of marine life, and

Ponting's photograph of the headquarters for the expedition on Cape Evans shows the hut surrounded by skis, sleds, and crates of food and medicines.

A January scene in Antarctica showing the midsummer ice floes in the Ross Sea, against the backdrop of the shining mountains along the coast.

several investigators are convinced that dolphins, at least, have language and may be trained to communicate with man.[3]

Five hundred miles to the east, Amundsen's *Fram* lay at anchor in the Bay of Whales, on the eastern edge of the Ross Ice Shelf. His men quickly set up a prefabricated hut, storerooms, and workshops, cutting some of the storage rooms out of the ice itself, as had been done in the Arctic.

During the waning days of the Antarctic summer of 1910–1911, Amundsen and his men set up a chain of food caches at the 80th, 81st, and 82nd parallels. He hoped to make his dash for the pole in March 1911. But this was not to be. Like Scott, whose men also were laying in advanced supply depots, Amundsen had to wait out the oncoming Antarctic winter.

Impatient, the Norwegians sallied out of their stronghold, Framheim (Home of the *Fram*), early in September, but were quickly

On June 6, 1911, Scott's forty-third birthday was observed with a sumptuous dinner that included seal soup, roast mutton, and liqueurs. This photograph, taken by Ponting, shows Scott seated at the head of the table.

driven back by the violence of the spring storms. On October 19, 1911, the Norwegian party began to move out along the ice shelf toward the mountains and the polar plateau. The weather had moderated and the warm sunshine brought a hint of spring to the frosty air.

Amundsen's expedition was organized down to the last detail. There were four sleds, each pulled by thirteen Greenland dogs, each with a carrying capacity of eight hundred pounds of supplies. There was enough food and cooking fuel for four months.

At the outset, the days were bright and the cold was not severe. The Norwegians drove their yelping dogs with skill and authority, making strong headway south. They crossed the shelf rapidly. Only when they reached the polar mountains did the journey become difficult.

Slowly up the mountain barrier they climbed, through a network of glaciers laced with horrendous blue crevasses. Sometimes the way was so slippery that neither man nor dog could maintain footing. Frequently they ran into blind alleys and had to backtrack. They made numerous reconnaissances to find through routes, and blizzards pinned them down for days.

When they reached the hard, wind-roughened ice of the plateau, the men dubbed the region "The Devil's Dance Hall." On December 13, 1911, they reached latitude 89 degrees, 37 minutes south and there made camp. In his tent, Amundsen wrote, "I awoke several times possessed of the same feeling I recall having had as a small boy on Christmas Eve—tense expectancy of what was about to happen."

At 3 p.m. the next day, it happened. The Norwegians stood on the geographic south pole. At this windswept and desolate spot all five men of the party grasped the Norwegian flag and drove it firmly into the ice. "So we plant you, dear flag of ours, on the South Pole and name this tract of land on which we stand, King Haakon VII's Plateau," they wrote in a note planted with the flag. The message was signed by Olav Bjaaland, Helmer Hansen, Sverre Hassel, Oskar Wisting, and Roald Amundsen. At the time Amundsen was thirty-nine years old.

The return to Framheim proved relatively easy. The party reached quarters at the Bay of Whales on January 25, 1912, bringing back only two sleds and eleven dogs, of the four sleds and fifty-two dogs they had started with. They had traversed a round-trip distance of 1900 miles in 99 days.

Victory and death

On the other side of the Ross Sea, meanwhile, Scott went ahead with his preparations for the traverse to the pole, which was some 60 miles farther by air from Cape Evans than from the Bay of Whales. A crew sailing the *Terra Nova* across the Ross Sea had located the *Fram*. Scott was disturbed that Amundsen had camped two or three days' journey nearer the pole than he. For even on a trek of some 700 to 800 miles, 60 miles in the Antarctic was no small matter. For exhausted men, it could mean the difference between life and death.

On the other hand, Amundsen knew little about the routes through the mountains. Scott's route had already been trail-blazed by Shackleton. It led across the ice shelf and up the giant Beardmore Glacier, the "royal road" to the pole.

What a road the Beardmore is! This river of ice is 100 miles long. It flows from the 9000-foot plateau down to the shelf, which is at about sea level, or a little above. Where it joins the shelf, the Beardmore is 30 miles wide. The volume of water in this frozen torrent can be compared with that in the Amazon River.

Having laid in a system of advance food depots, Scott and his party started off on their journey November 2, just 14 days after Amundsen's sleds moved out of Framheim 500 miles to the east. The route Scott took was 922 miles long, from Cape Evans to the pole. It was probably the most tortuous journey a group of men has ever attempted.

Not knowing how Amundsen was faring, Scott spurred himself and his party by racing Shackleton's record of 1907, and actually beat Shackleton's time to the top of the Beardmore Glacier. Then summer storms delayed his party, which made little progress across the plateau before the end of December. On January 3, 1912, Scott estimated that he was 150 miles from the pole. He chose four men to accompany him on the last lap and sent the others back to McMurdo Sound.

Those who went forward with him to glory and to death were Dr. Wilson, thirty-nine years old, physician, zoologist, and chief scientist of the expedition; Petty Officer Edgar Evans, thirty-seven; Captain Lawrence E. G. Oates, thirty-one, of the Enniskillen Dragoons; and Lieutenant Henry R. Bowers, twenty-nine, of the Royal Indian Marines.

Soon there came a day on the plateau when the men faced the

An aerial view looking east over the Beardmore Glacier, taken at an altitude of 20,000 feet above sea level by a Navy photo-mapping aircraft.

A photograph by Ponting shows the Siberian ponies and dogs awaiting trial as draft animals to haul Scott's sleds from Cape Evans to the pole.

Another Ponting photograph reveals Captain Lawrence Oates (right) and another expedition member seated in the hut at Cape Evans, late in 1911.

vast desert of ice with only their muscle power for transportation. The Siberian dogs and ponies they had been using had succumbed. Most had died of exhaustion on the Beardmore. Scott's polar party now had to haul their own sleds, which was the beginning of the end for them. Amundsen's great advantage was the use of Greenland dogs. Trained on the Greenland ice cap, these animals could pull a sled laden with half a ton of supplies fifty miles a day. That is a good pace for modern gasoline-powered tractor trains on the ice today. The Greenland dogs were adapted for ice; the Siberian dogs were not.

According to other members of the *Terra Nova* expedition, Scott had attempted to purchase Greenland dogs, but the Danish authorities had declined to sell them to him.[4] The reasons for this are not clear. In any event, the Asian dogs and ponies proved quite unequal to the Antarctic ice cap. When they had perished, there remained only the minds and muscles of five Englishmen to conquer the pole.

It was a bitter summer on the plateau that year. The men struggled into the polar wind, bowed, dragging the sleds behind them, sometimes pushing them ahead, in temperatures of 30 degrees below zero Fahrenheit. On the evening of January 15, 1912, with the sun shining brilliantly on the ice, Scott noted in his diary, "Height 9020 feet, temperature —25°F. It is wonderful to think that two long marches should land us at the pole." The diary continues:

Jan. 16: Height 9076 feet. Temperature −23.5°F. We started off in high spirits this afternoon feeling that tomorrow would see us at our destination. About the second hour of the march, Bowers' sharp eyes detected what he thought was a cairn. We marched on and found it was a black flag tied to a sledge bearer; nearby, the remains of a camp, sledge tracks and ski tracks going and coming, and the clear trace of dogs' paws . . . many tracks. The Norwegians have forestalled us and are first at the pole! It is a terrible disappointment.

Scott and his companions reached the pole the next day, January 17, 1912. England had lost the race.

The pole yes, but under very different circumstances from those expected. We have had a horrible day. Great God! This is an awful place and terrible enough for us to have labored to it without the reward of priority.

Well, it is something to have got here and the wind may be our friend tomorrow. Now for the run home [Cape Evans] and a desperate struggle. I wonder if we can do it.

Evans was the first to perish. After a fall, which might have given him brain concussion, he grew weaker day by day. He finally collapsed and died on the descent of the Beardmore Glacier. The four men struggled on, weak now, making only five or six miles a day.

In mid-February Oates's feet became frostbitten. Gangrene set in. He knew he was doomed, but he kept going. Every step was agony for him. Repeatedly he begged Scott to leave him behind, but Scott refused.

Sick and weak, the men reached the shelf, where the going should have been easier. But without warning a nine-day blizzard pinned them down only eleven miles from a supply depot which they could not reach.

While the blizzard raged at its height, Oates limped out of the tent, saying he did not expect to be back soon. He vanished into the swirling snow and did not return. Scott, Wilson, and Bowers lay in their sleeping bags, wondering if any of them would be alive when the blizzard ended.

On March 29, 1912, Scott made the last entry in his diary:

Every day we have been ready to start for our depot 11 miles away, but outside of the door of the tent it remains a whirling drift. I don't think we can hope for any better things now. We shall stick it out to the end, but we are getting weaker, of course, and the end cannot be far. It seems a pity but I do not think I can write any more. R. Scott. For God's sake look after our people.

So ended the race for the south pole.

Eight months later a search party, led by the surgeon Dr. E. L. Atkinson, found the tent. Wilson and Bowers were wrapped in their sleeping bags. Scott had thrown back the flaps of his bag and had opened his coat. At the time of his death he was forty-three years old.

Thin ice or thick

The most important scientific results of the journey, which was not wholly in vain, were thirty-five pounds of geological specimens on a half-buried sled near the death tent. Wilson had insisted on dragging them back.

Among the samples was coal, found in thin layers in the polar mountains. In the sandstones which bordered the Beardmore Glacier there were fossils of ancient plants. These included a big-leafed fern called *Glossopteris,* fossil wood, and winged grains of pollen.

Coal had been found, too, by members of Shackleton's party, but Wilson's collection was the largest and most complete array of specimens yet extracted from the polar mountains. Here was evidence that at a time past, this region within 150 miles of the south pole had been warm enough to grow coal-making vegetation. So far as man's knowledge about his environment was concerned, this was the most significant Antarctic discovery since Ross's day.

The fossils told a strange story, indicating not only that this land of seemingly eternal ice had once been temperate or tropical, but that its relationship to the sun may have changed. The coal and the fossil wood suggested the possibility that the continent had moved or drifted from a geographical position farther to the north, since the growth of plants luxuriant enough to have become metamorphosed into coal might not have developed in a region where the night is six months long.

There is by no means general agreement, however, that these conditions of polar sunlight would inhibit widespread plant growth. A short spring and summer growing season, when the sun is shining twenty-four hours a day, might produce a substantial vegetation if other conditions, temperature and moisture, were favorable.

In any case, the fossils that the heroic party struggled so desperately to show to the world posed the central question in the multifaceted Antarctic mystery. Once the land had been open to the sun, without ice. Then an ice age had come. What processes in alteration of climate or even in the evolution of the earth did the change represent? That essentially was the legacy Scott and his companions bequeathed to the world.

In 1913 a twelve-foot cross of Australian jarrah wood was erected atop Observation Hill, which overlooks the present Naval Air Facility on Ross Island and the old *Discovery* hut nearby. Etched with infinite care on the upper limb of the cross are the names of the polar party: Scott, Wilson, Oates, Bowers, and Evans. Beneath is carved a line from Alfred, Lord Tennyson's "Ulysses": "To strive, to seek, to find and not to yield."

The scientific results of the expedition began appearing in 1914, after the return of the *Terra Nova* to England, but because of World War I the full account of the scientific findings was not published until 1922. The geophysical studies contributed to the validity of Murray's "hypothetical continent." The continental nature of Victoria Land was established, but whether other areas beneath the ice cap were a part of the continental mass remained in doubt.

Left: *The jarrah-wood cross on Observation Hill, erected in 1913 to the memory of Captain Robert Falcon Scott and his party.* Below: *Two views of the bunkroom in Scott's hut at Cape Evans, after it was restored in 1961 by the New Zealand government. The personal clothing hanging on the bunks was left by the men when the hut was abandoned.*

One of the specific goals laid upon the expedition was to determine the extent and thickness of the ice cap. This had an important bearing on a number of planetary questions, including a theory of ice ages and the planet's water budget. It was also important to know whether the ice was advancing or retreating. This too had a direct bearing on theories of ice ages and possibly, as had been suggested in a tentative way, it might affect sea level.

The consensus of the *Terra Nova* expedition scientists was that, compared to the mile-thick ice sheets that overlay the northern hemisphere during Pleistocene glaciations, the ice cap of Antarctica was a thin one. The idea of thin ice in Antarctica was perhaps the most influential conclusion of the expedition. It dominated theories of glaciation or glacierization* for the next forty years.

Scott estimated the area of the ice sheet as two-thirds of a circle with a radius of 1200 miles, or barely 2,000,000 square miles. As we now know, this was less than one-half the 5,500,000 square miles actually covered by the ice cap. "We may surmise," Scott wrote, "50 to 100 yards movement per annum across the edge of the [polar] plateau, with a thickness of 700 to 800 feet."

Scott realized his estimates were made from a limited view of the region and without soundings. But there was no means of sounding the ice in 1912. He knew he was guessing. "I am only constructing an edifice of theoretical bricks," he said. "This is merely a fine effort of the imagination."

The thin-ice concept was supported by all the *Terra Nova* expedition's physical scientists. Griffith Taylor, the leading geologist with the expedition, stated in his résumé of the physiography and glacial geology of Victoria Land, "The comparatively slight depth of the outlet glaciers seems to indicate that the ice cap is not very thick, probably one or two thousand feet only."[5] To this Charles S. Wright and Raymond Priestley, physicists with the expedition, added that "the depth of the ice appears nowhere greater than 2000 feet, except at the head of glaciers."[6]

The fact that the ice cap was in motion was established by observation, but the rate of movement from the polar plateau was guesswork. Actual measurements on the leading edge of the Ross Ice Shelf, however, showed that it had receded an average of 15 to 20 miles from 1840, when Ross measured its northern boundary, to

* Some geologists prefer the term "glacierization" when they refer to the development of an ice sheet. The term was proposed in 1922 by the physicists Charles S. Wright and Raymond Priestley, who used the term "glaciation" to mean land eroded by ice.

1902, when Scott fixed the position of the barrier. At one point, the shelf had receded southward a total of 45 miles. Between 1902 and 1911, however, no further recession had been noted.[7]

Charles (later Sir Charles) Wright did a great deal of work on the shelf. He estimated it was moving to the east-northeast at the rate of 500 yards a year. On the basis of measurements made by the *Terra Nova,* Wright reported that the shelf was 400 miles wide from east to west and 600 feet thick. He assumed the shelf moved under its own weight, an idea which seemed to overlook the possibility that this enormous piece of floating ice, the size of Spain, was an extrusion over the Ross Sea of the continental ice sheet.

Of the land, the geologist Taylor concluded that the mountain range (the Trans-Antarctics) running due north and south from McMurdo Sound to the polar plateau constituted a continental shoreline, with a steep eastern drop, but a more gradual slope to the west. Taylor, who was Australian, was impressed by the similarity of this apparent Antarctic structure to that of Australia. He theorized that movements within the earth during Tertiary times, 1,000,000 to 63,000,000 years ago, a period of continent building, had tilted both the land masses of Antarctica and Australia in the same way, depressing the shoreline in the east and elevating it in the west.

Returning to Antarctica in 1960, Sir Charles Wright examines the ice crystals formed on the photographic equipment used by Herbert G. Ponting in 1911 when they were both members of Scott's final expedition.

The land below the central ice plateau of Antarctica, Taylor suggested, was a peneplain—land planed down nearly table-top flat by erosion. It was similar, he thought, to the plains of western Australia or the Great Plains of North America. The fossils provided the evidence that warmer conditions prevailed in Tertiary times, he noted.

The expedition's assistant geologist Frank Debenham, who has written extensively and authoritatively about Antarctica, defined the foundation rocks in Victoria Land as gneisses, schists, quartzite, and crystalline limestones of Pre-Cambrian age, more than 600,000,000 years old. In earlier times, before these mountains were uplifted, the land was covered by the sea. The limestones testified to this. So did the thick layers of sandstones, whose grains were water polished. Now the sandstone layers were nearly three miles above sea level, mute witness to the convulsive changes within the earth.

In a portion of the mountain chain called the Royal Society Range, the sandstone grains showed evidence of having been worn by winds and there were fossils of land and fresh-water plants. Debenham postulated that at one time the region had been a low-lying desert of sand and dunes, perhaps representing another era in the long past of the region, after the sea had gone. Debenham visualized a semi-arid climate without a dense plant or animal population. There were no fossil evidences of extensive plant life in the Royal Society sandstones, and animal fossil remains found elsewhere in Antarctic sandstones have been limited to small marine invertebrates. According to Debenham, worm markings, ripple marks, and the casts of sun cracks indicated conditions such as now exist in the Gobi desert.

In the vicinity of the Beardmore Glacier, some 450 miles to the south, the fossils told another story of climate ages past. When these fossilized plants were growing, the climate was humid and the landscape marshy. Debenham surmised that there were rapidly flowing, but not large, rivers throughout the region. He noted that in places the sandstones contained small pockets of conglomerate, bits and pieces of rocks glued together after having been brought into contact by the wash of running water.

The medieval speculators had been right, in one sense. There had been, in the south, a fair land, but only in the long ago. Man, it seemed, had arrived several hundred million years too late.

Except for lichens, struggling to make a living on the barren rocks, and strange, wingless insects that appeared in the coastal regions adrift on the winds, there was no life on the land. Only in the seas did native animal life exist. Rich in dissolved oxygen, Antarctic

Boulders, eroded by blowing sand and ice crystals, litter the dry floor of Wright Valley in the Royal Society Mountains, west of McMurdo Sound.

waters support vast crops of plankton which is harvested by huge whales. The food chains of the southern ocean provide well for seals, for penguins, and for birds of the air. Of native land animals, however, there is no sign, nor so far any indication that land vertebrates ever existed in Antarctica.

One continent or two

After the conquest of the pole, the next major physical challenge in the Antarctic was the crossing of the ice cap from the Weddell

Sea, on the Atlantic side of the continent, to the Ross Sea, on the Pacific side. Such a project had strong scientific justification in the growing belief that a third sea, or an arm of one, existed in the 1800-mile expanse between the two embayments.

The crossing of Antarctica, therefore, became the new goal. It was considerably more difficult than simply driving for the pole from a well-stocked base. It required two widely separated bases, one in the Weddell Sea region opposite South America, and the other in the Ross Sea region, opposite New Zealand.

The project was attempted by the German explorer-scientist Dr. Wilhelm Filchner in 1912, and by Shackleton in 1914. Both failed because their ships became beset in the treacherous ice of the Weddell Sea. Filchner's ship, the *Deutschland,* was extricated after an ice-imprisoned drift of nine months, but Shackleton's ship, the *Endurance,* was ultimately crushed and sunk.

Fortunately, the demise of the *Endurance* was protracted enough to enable Shackleton and his men to unload their stores onto the ice. Having stocked three of the ship's longboats with as many supplies as they could hold, Shackleton's party pushed and hauled the boats across the ice, sailing them when open leads of water appeared.

They managed to reach Elephant Island in the South Shetlands. Leaving 22 men on the island, Shackleton and a crew of five sailed one of the boats in search of help across 800 miles of the Scotia Sea to South Georgia. There was a coaling station on the island, but the exhausted men beached their craft on the opposite side and had to climb a 9000-foot mountain range to reach it. At the end of August 1916 the castaways on Elephant Island were rescued by a Chilean ship which responded to Shackleton's pleas for help.

This incredible tale of Antarctic survival was drowned out by the cannon of World War I. It is remarkable that Great Britain supported such an expedition while the country was locked in that conflict.

The abortive efforts by Filchner and Shackleton were the last seeking to cross the ice cap over the snow until the International Geophysical Year (1957–1958). Explorers took to the air in pursuit of the problem of whether there existed an arm of the sea between the Ross and Weddell embayments. Sir Hubert Wilkins, a pioneer in polar aviation, attempted to reconnoiter the ice cap between the Panhandle and the Ross Sea by airplane in 1928–1929 and again in 1929–1930. Each time bad weather forced him to abandon the effort.

Rear Admiral Richard E. Byrd, who led the United States back into the Antarctic in 1928 after a hiatus of nearly ninety years, used aircraft extensively to survey the West Antarctic interior from his camp, Little America, near Amundsen's old base on the Ross Ice Shelf at the Bay of Whales. In one of these sorties he found a coastal range of volcanic mountains in the region he named Marie Byrd Land. It comprises the third mountain system in Antarctica.

At the end of November 1929, Byrd, Bernt Balchen, A. C. Mc-Kinley, and H. June flew from Little America to the south pole and back, a round trip of 1500 air miles. Byrd relied on aircraft extensively for reconnaissance on his second expedition in 1933–1935 and on his third, in 1939–1940 (the Antarctic Service Expedition).

In 1935 Lincoln Ellsworth, with H. Hollick Kenyon as pilot, flew from Dundee Island near the Panhandle across West Antarctica to the Bay of Whales, making three landings on the ice en route. A German expedition under Alfred Ritscher surveyed Queen Maud Land on the Atlantic side of the continent from the air in 1938–1939. There was some speculation that Hitler was interested in developing a submarine base in the region, from which the undersea boats could harry shipping forced to detour around South America, in the event the Panama Canal was knocked out. Nothing came of this supposed project, however.

While the airplane was used with fair success as a reconnaissance vehicle, a quarter of a century was to elapse after Wilkins' flights before aircraft became reliable enough for extensive Antarctic transportation.

Visual inspection of the ice cap from the air was useful in finding gross features of the landscape and plotting them on maps. But it shed little light on the basic questions of ice thickness and the structure of the land beneath. These questions had to be answered by expeditions wending their way across the surface.

Griffith Taylor and Raymond Priestley had suggested that the Ross and Weddell embayments were the extremities of an 1800-mile graben, or sunken region, which ran all the way across Antarctica. If this hypothetical sunken region, called the Great Antarctic Trough, did exist, it was likely that the Ross and Weddell Seas were connected by water. The flights of Wilkins, Byrd, and Ellsworth found no sign of open water in West Antarctica. But these observations did not eliminate the possibility that a great trough might exist under the ice and be invisible from the air.

The hypothesis of a trough was based on geological clues that two land masses, not one, existed in Antarctica. One clue was that the structure of the mountains on the Panhandle is different from that of the Trans-Antarctic chain bordering Victoria Land and the polar plateau. In the Panhandle, beginning 550 miles south of Patagonia, the mountains structurally are akin to the Andes Mountains of South America, characterized by folded sedimentary and volcanic rocks of Mesozoic age, 63,000,000 to 200,000,000 years old. On the margin of the Ross Sea, some 1200 miles to the west, the Trans-Antarctic Mountains are formed of upraised blocks of Pre-Cambrian rocks, at least 600,000,000 years old.

Since the Panhandle mountains were found to be similar in structure and age to the Andes, they are frequently referred to as the "Antarctandes," a term applied by Henryk Arctowski, a Polish geologist, on the *Belgica* expedition of 1898. They thus formed a part of the great Andean, or Cordilleran, fold. This feature of the planet stretches from Alaska through the Americas to the tip of Patagonia. It continues, submerged, in a lazy arc through the Drake Passage of the South Atlantic Ocean. Then the mountains arise out of the sea, steep and ice-clad, on the Antarctic Peninsula. The peaks of the Panhandle apparently are part of the same crustal fold that one sees in the Rocky Mountains from the streets of Denver.

The Panhandle thus seemed to be more closely related to South America, from which it was separated by the ocean, than to continental Antarctica, to which it was joined by the ice sheet. Or were these land systems joined? Did a sub-sealevel trough actually separate them? Priestley commented, "Morphologically, Graham Land [the Panhandle] stands as a mirror image of Patagonia across the deep water of the Drake Strait. The geologist is thus faced with the question: 'Am I dealing with one continent or two?'"

Allied to this question was the problem of ice depth. If the estimates of Taylor, Priestley, and Wright were right about the ice's being relatively thin, no thicker than 2000 feet, the Antarctic cap would be exceptional rather than typical of continental ice sheets. The depth of northern hemisphere ice sheets of the Pleistocene is estimated to have been 10,000 to 12,000 feet at the centers from which they radiated southward. If the Antarctic cap was exceptional, then it appeared reasonable to assume it was in a state of transition, and Scott's observations of ice retreat in McMurdo Sound, as well as the retreat of the Ross Shelf front between 1840 and 1902, seemed to confirm this.

The notion that the Antarctic cap was thin was supported by

visual observation that its contour generally seemed to follow that of the land beneath. Ice that revealed land contour could not be very thick.

The estimate of thin ice had a direct relationship to a theory of ice ages, which had evolved late in the nineteenth century and now seemed to be confirmed by observations in Antarctica. The theory held that glacierization was caused by astrophysical events affecting the earth's climate. These might be changes in the energy output of the sun or the planet's passage through a cloud of dust which would reduce the amount of solar radiation falling on the earth. Even a small decrease in sunshine would result in significant cooling in polar regions, which receive the least amount of the sun's heat because of the wide angle at which its radiation strikes the earth in the high latitudes, and which have a high "albedo," the ability to reflect radiation back into space.

If such changes in solar radiation did occur, it was reasonable to expect their effects on the climates of the northern and southern hemispheres would be about the same. Ice ages, therefore, would tend to come and go simultaneously in both hemispheres.

This theory of simultaneous and synchronous ice ages was supported by geological evidence that parts of South America, New Zealand, and Australia were glacierized at about the same time as North America, Europe, and western Siberia, although not to the same extent. If that was so, the ice would have retreated in the southern hemisphere at about the same time as it did in the northern hemisphere.

How, then, did one account for Antarctica? Here was an ice age existing in the south at a time when only the residue of one remained in the north. The only matching feature in the northern hemisphere was the Greenland ice cap, hardly a pint to Antarctica's gallon.

Geographically, of course, the Arctic and Antarctic are entirely different. They are far from the mirror images their names suggest. The north pole is in the Arctic Ocean, which is surrounded by continents, while the south pole is in a continent surrounded by oceans.

Which arrangement makes for a colder climate? Experts cannot agree. The Antarctic is much colder and has more ice than the Arctic. Is it colder because it has more ice? Or does it have more ice because it is colder? Scott contended that the ice would become thicker if a warming trend developed, and would diminish in very cold periods. He reasoned that as warm air held more moisture than cold air, during warmer periods there would be more snow.

In any event, if Antarctica with its huge ice cap was to fit into an astrophysical theory of synchronous, bi-hemispheric ice ages it had to be a residue of something much greater, like Greenland's ice sheet. But it was stretching reason to think of a multimillion-square-mile ice cap as a mere leftover. Therefore, to satisfy the astrophysical theory, the Antarctic ice cap had to fulfill two conditions: it had to be thin, and it must have waned. This seemed to be precisely the case. The idea of a thin ice cap became firmly established in the literature of ice ages. A. P. Coleman, professor of geology at the University of Toronto, wrote in 1926: "It appears, therefore, that the Antarctic ice sheet, though of enormous extent, may actually contain less ice than any one of the three great Pleistocene ice sheets in the Northern Hemisphere."[8]

In another standard text, Richard Foster Flint, professor of geology at Yale University, wrote in 1945: "It is unlikely that the ice thickness exceeds 2000 feet except very locally; probably its average thickness is considerably less."[9] On the basis of magnetic declination, T. W. Edgeworth David, professor of geology at the University of Sydney, Australia, concluded that the ice on the high plateau of East Antarctica at the magnetic south pole "is of no very great thickness."[10]

Other apparent indications of thin ice were provided by aerial photographs by Ellsworth on his 1935 flight across Marie Byrd Land in West Antarctica. From studies of the pictures it was concluded that "the interior of the ice cap . . . does not completely mask the underlying topographical features."[11] In 1940 Dr. Laurence M. Gould, a veteran polar geologist, suggested that the structure of the continental margin about the head and western edge of the Ross Sea "is not one that leads to the conception of a continental ice of great thickness."[12] There are other confirmations of the thin-ice concept. In the absence of actual soundings of the ice, the visible clues all pointed to it.

Against the weight of authority and available evidence, however, one investigator offered contradictory data. Professor Robert L. Nichols, head of the geology department at Tufts University, suggested in 1949 that the unusual depth of the Antarctic continental shelf indicated that the continent was being depressed by a great load of ice which might be several miles in thickness. The break in the slope of the Antarctic continental shelf (the depth at which it plunges down to sea bottom) had been ascertained to be four times the average for continents. This could be attributed reasonably to the depression of the land by the ice load.[13] Following this assumption, Nichols computed the thickness of the ice load mathematically and estimated that the average thickness of the continental ice sheet ranged from 6300 to 8500 feet.

Was the ice cap thick or thin? Did it cover one continent or two? By the middle of the twentieth century, 110 years after Ross sailed into McMurdo Sound, these questions remained in the foreground of man's investigation of the earth.

Antarctica's secrets were not to be yielded up to single-nation expeditions. The physiographic problems of the region required a multi-nation attack and an advanced technology capable of supporting investigators in the field with icebreaker ships, aircraft, and tractors. During the period between the World Wars neither the motivation nor the technology was available for this kind of effort. Yet the challenge was by no means ignored.

The British remained active in Antarctic waters. In the 1920s there was a series of voyages in which Scott's famous old vessel, the *Discovery*, was used in Antarctic oceanography. In 1929 the ship was succeeded by the *Discovery II*, which continued a sequence of annual cruises in the southern ocean until the outbreak of World War II. After the war these research cruises were resumed and in 1954 the British Admiralty published a monumental survey of the southern ocean, comprising 130 reports.

The Norwegians were active too. Between 1927 and 1937 eight voyages to Antarctic regions were promoted by a Norwegian whaling magnate, Lars Christensen. Aerial surveys were made of Enderby Land in the African quadrant of Antarctica. Much of Queen Maud Land in that quadrant was photographed from airplanes—the first use of a technique of photocartography which was to be used extensively in the 1960s.

Until 1934 the relationship of the Panhandle and the main ice sheet remained uncertain. Wilkins' flights at the base of the peninsula had provided evidence that it was an archipelago of mountainous islands rather than a continental extrusion. Between 1934 and 1937, however, the British Graham Land expedition, led by John Rymill, established that the Panhandle was connected to the mainland ice sheet.

Even during World War II the British remained active in the region. In 1943–1944 survey parties built a group of stations in the South Orkney and South Shetland islands and on the Panhandle, partly for scientific purposes and partly as a means of preserving a claim to the islands and to a large pie-slice of Antarctica called the Falkland Islands Dependencies. The stations were manned for a decade. The first continuous weather observations were made there—the beginning of systematic Antarctic meteorology on a local scale.

Argentina began sending regular expeditions into the Panhandle in 1942, and Chile followed in 1947.

So far as the United States was concerned, Byrd was the principal figure in the Antarctic during the period between the wars. After World War II the United States Navy mounted Operation High Jump, the largest sortie into Antarctic regions ever made. Byrd directed the operation, which included a fleet of thirteen ships under the command of Rear Admiral Richard H. Cruzen. Nearly four thousand men traveled to the Antarctic on this expedition. News-media writers and photographers were taken along for the first time. Extensive aerial surveys of the lower Panhandle, Marie Byrd Land, and the East Antarctic ice sheet were made successfully.

Between 1946 and 1948 the private expedition of Finn Ronne, a commander in the United States Naval Reserve, reoccupied the east base of Byrd's Antarctic Service Expedition in the Panhandle. It was on this expedition that Dr. Nichols of Tufts University developed his thesis of thick ice, which ran contrary to the prevailing view at the time that the ice sheet was a thin one.

Ronne, Nichols, and others made an 1180-mile sled journey on the east coast of the Panhandle, spending 105 days in the field. With the expedition were two women, Mrs. Edith Ronne and Mrs. Jennie Darlington, wife of one of the expedition's aircraft pilots. They were the first women to winter over in Antarctica.

The birth of the IGY

On the evening of April 5, 1950, a notable group of scientists attended a dinner party at the Silver Spring, Maryland, home of James A. Van Allen, for whom the radiation belts were later named. The guest of honor was Sydney Chapman, an English geophysicist.

At this gathering was born the International Geophysical Year (IGY). It was suggested by Dr. Lloyd V. Berkner, who had been a radio engineer on the first Byrd Antarctic expedition, as a third international polar year. The first polar year had been staged in 1882–1883, when eleven nations dispatched expeditions to the Arctic. The second was held in 1932–1933, organized largely by the International Union of Geodesy and Geophysics, which had been founded in 1919. In this way, a tradition was forming to hold an international polar year every half-century. But neither Berkner nor anyone else at

Van Allen's party wanted to wait that long, according to an account of that evening related by the Canadian geophysicist J. Tuzo Wilson.

Advances in technology during and after World War II had produced reliable tractors capable of traveling long distances over ice in a rigorous climate. Also longer-range, more reliable aircraft existed, capable of supporting parties at considerable distances from base. Now was the time to tackle the Antarctic.

Moreover, a third international polar year would be able to put to good use the new developments in instruments. The idea was broadened to include simultaneous observations of geophysical phenomena over the entire planet—on the land, in the seas, and in the atmosphere. What science had lacked in meteorology, in magnetism, in ionospheric physics, in oceanography were observations made in a great number of different places on the planet by different people —all at the same time and with the same objectives.

The proposal for such global effort was submitted to the International Council of Scientific Unions (ICSU), the parent body of international organizations in such fields as physics, chemistry, geography, geophysics, radio, and astronomy. The ICSU set up a committee to organize the IGY, the Comité Spécial de l'Année Géophysique Internationale (CSAGI), which held its first plenary session at Brussels in July 1953. A provisional program was drawn, based on proposals from twenty-three nations.

For the Antarctica project, the major geophysical questions were defined as the volume and structure of the ice sheet, the nature of the land beneath the ice, the influence of the ice cap on global and southern-hemisphere weather, the aurora, and the high atmosphere, with emphasis on the ionosphere. For example, what happened to the ionosphere, an electrified region of the upper atmosphere which supposedly formed only in sunlight, during the long polar night?

The IGY evolved swiftly to become the most successful cooperative effort in the history of nations. It was also man's greatest effort to unravel the mysteries of his environment. Directly involved in making observations were 67 nations, which sent 12,000 scientists to 2500 stations, including posts at the geographic south pole and on ice islands in the Arctic Ocean near the north pole.

Operation High Jump had paved the way for the grand invasion of Antarctica by 12 nations, which tackled this roughest of all terrestrial environments. Argentina, Australia, Belgium, Chile, France, Japan, New Zealand, Norway, South Africa, the Soviet Union, the United Kingdom, and the United States agreed to set up 60 stations

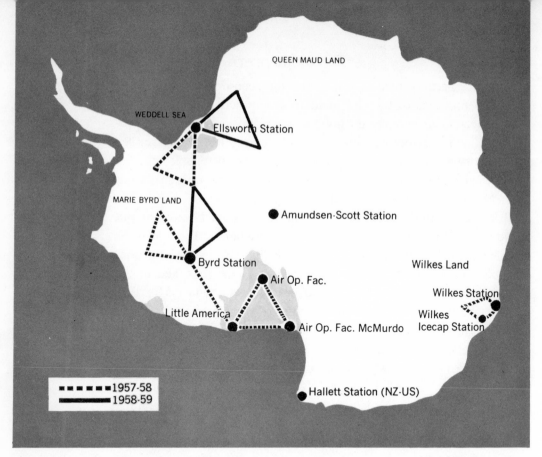

The map shows the following labels:

QUEEN MAUD LAND

WEDDELL SEA

Ellsworth Station

MARIE BYRD LAND

Amundsen-Scott Station

Byrd Station

Air Op. Fac.

Wilkes Land

Wilkes Station

Little America

Wilkes Icecap Station

Air Op. Fac. McMurdo

Hallett Station (NZ-US)

■ ■ ■ ■ ■ ■ 1957-58
━━━━━━ 1958-59

The Antarctic Program of the United States National Committee for the International Geophysical Year.

on the continental ice cap and on Antarctic and sub-Antarctic islands. Not to be outdone by the American decision to build a station at the south pole, the Russians proposed a station at the pole of inaccessibility—the point on the ice cap most distant from the coasts.

Admiral Byrd was honorary chairman of the United States Antarctic Committee. Its chairman was Dr. Gould, later professor of geology at the University of Arizona, who had been chief scientist on Byrd's 1928–1930 expedition. The United States Navy was commissioned to support the scientists on the ice and immediately began to construct stations. Even before the IGY started, the region known as the Ross Sea Dependency of New Zealand became a *de facto* Navy preserve.

Two years before the start of the IGY, which opened formally July 1, 1957, the Navy began moving Seabee construction crews and matériel into Antarctica. The logistics required at least a year's advance planning. Bulk material must be shipped a year before it is scheduled to be used, for by the time the ships can move through

A construction crew at Little America Station, in early November before the snow has melted.

Seabees laying the foundation for a hut in the midsummer sun at Hallett Station on Cape Adare.

the ice of the Ross Sea the summer is usually well advanced and little outdoor working time remains before the autumn winds rise and paralyzing cold arrives. The outdoor working season is limited from about the beginning of October to the middle or end of February. In Antarctic construction, you must plan ahead.

Historically, the Navy operations in support of the IGY were authorized on September 17, 1954. The following December 1, the icebreaker USS *Atka* got under way from Boston for Antarctica, to scout for bases. Admiral Byrd was appointed officer in charge and Captain George J. Dufek (later Rear Admiral), acting commander of the Antarctic Support Force, which became designated as Task Force 43. In stanch military tradition, the support activity was coded Operation Deep Freeze and the 1955–1956 construction period was Deep Freeze I.

From the time the icebreaker USS *Glacier* crunched her way through the ice into McMurdo Sound and put ashore, in December 1955, an airstrip-survey party to commence the construction of the Naval Air Facility, until mid-1958, when the American bases were established, a complete metamorphosis was achieved in the support

Left: *The icebreaker USS Glaci* *plowing through the bay ice of M* *Murdo Sound.* Above: *Unloading car* *from a supply ship onto sleds, to* *pulled by tractors to McMurdo Statio* *located eight miles away.* Right: *Nig* *scene of the main street at NAF M* *Murdo.*

of polar science. Instead of a single camp, lonely and isolated, a continental array of bases arose on the ice cap, linked by radio and aircraft as well as by a common purpose.

The Seven Cities of the seventh continent

Between the entrance of the Ross Sea and the Filchner Ice Shelf, the United States built seven stations. Hallett Station arose on Cape Adare, hard by a rookery of thousands of penguins nesting in the lee of the mountain wall guarding the ice sheet of Victoria Land. Southward, the Naval Air Facility, a geometric pattern of semi-tubular Jamesway huts insulated with fiberglass, and boxlike plywood Butler Buildings, painted orange, was built on Ross Island in the shadow of Mount Erebus, the snow-covered volcano. Naval Air Facility McMurdo, with streets known as Honeybucket Lane and quarters called the Hard-Luck Hilton, evolved less than a mile from Scott's 1901–1904 quarters at Hut Point. NAF McMurdo served both as a science station and as port of entry, including an airport

scraped out of the bay ice of the sound, called Williams Field. For
the edification of visitors, Seabees painted a large chamber-of-com-
merce-style sign reading: WELCOME TO MCMURDO SOUND. PARA-
DISE OF THE ANTARCTIC.

Eastward across the Ross Shelf, Little America V was built on
an indentation called Kainan Bay. This last of the famous series
of stations founded by Byrd was east of the old sites on the Bay of
Whales. The 1954 reconnaissance of the *Atka* had discovered that
the Bay of Whales—a landmark since Amundsen's time—had disappeared.

More than 3000 men, 200 aircraft, and 300 vehicles were involved in the construction of the "Seven Cities" of Antarctica. A
fleet of cargo ships hauled the materials down to the ice continent
in the wake of icebreakers. Building materials for a station in Marie
Byrd Land on the West Antarctic ice sheet were unloaded at Little
America, and materials for a station at the south pole were put
ashore at NAF McMurdo. More than 1000 tons of matériel arrived
at these busy ports during Deep Freeze II (1956–1957).

A view overlooking the Jamesway huts of McMurdo Station, west across the frozen McMurdo Sound to the mountains of continental Victoria Land,

*about 58 miles away. The largest of all Antarctic stations, McMurdo
serves as the primary port of entry for United States ships and aircraft.*

On October 31, 1956, Dufek, who had been promoted to rear admiral and placed in command of Operation Deep Freeze, took off from McMurdo Sound in a Navy R4D aircraft with the fatalistic name *Que Sera Sera*. Five hours later he and his pilot, Lieutenant Commander Conrad Shinn, and members of the crew stepped out on the ice of the south pole—the first men to stand there since Scott and his party in 1912.

It was 58 degrees below zero Fahrenheit that day. Within a few minutes the telltale white patches of frostbite appeared on the cheeks of the men, between their sunglasses and the fur of their parka hoods. Dufek decided it was too cold to begin the construction of a station. Boosted off the polar ice with JATO bottles* the *Que Sera Sera* flew back to McMurdo.

* Because of the low atmospheric density, the use of JATO, or jet-assisted take-off, has been standard operating procedure for taking off from high plateaus in Antarctica.

The late Dr. Harry Wexler addressing IGY personnel at the historic dedication ceremony for the Amundsen-Scott South Pole Station.

The advance Navy construction force of eight men was flown to the pole in ski-equipped aircraft on November 19. In their wake lumbered big United States Air Force C-124 Globemasters, which were then the world's largest freight-carrying aircraft. While circling over the pole, the "Globies" parachuted prefabricated buildings, section by section. On November 26 a seven-ton tractor was successfully parachuted to the pole, and all hands turned to erect the station under the direction of Lieutenant John Tuck, military leader, and Dr. Paul Siple, scientific leader, who had been to Antarctica as an Eagle Scout with Byrd.

Back and forth the C-124s shuttled between the pole and Mc-Murdo, following Scott's old route over the gigantic Beardmore Glacier. A village of Jamesway huts arose at the southern axis of the earth, complete with a mosque-shaped radome and a forest of antenna masts. Surrounding this tiny community was an enormous pile of litter—broken crates, piles of cans, and assorted trash, which made the glamorous south pole look like the corner of a city dump and the Amundsen-Scott Station like some Hooverville of the 1930s. But there was no place to dispose of rubbish on the ice cap, and the crew simply followed the old Navy practice of tossing it "over the side." Eventually, of course, the pile was covered by a season's snow.

Elsewhere in Antarctica construction crews were busy too. The Seabees put up Wilkes Station on the Budd Coast of Wilkes Land, in the Australian quadrant of East Antarctica, and Ellsworth Station was built on the Filchner Ice Shelf on the Atlantic side. From Little America a team of Greenland-trained Army engineers, under Lieutenant Colonel Merle (Skip) Dawson, supported by Navy, Coast Guard, and Air Force personnel, laid a trail across the ice of West Antarctica to latitude 80 degrees south and longitude 120 degrees west in Marie Byrd Land, where Byrd Station was built. Construction material was hauled by tractor train over the trail, sometimes distinguished by the name of the Army-Navy Highway or US Minus 66. Supplies were also ferried by air from Little America and McMurdo and parachuted to the ice.

When the austral autumn came in March of 1958, the Seven Cities of Antarctica were more or less finished. At least they were habitable. In spite of their shacktown appearance, they represented one of the most difficult feats of logistics in the history of man. Said Admiral Dufek, "We had to spend some $245,000,000 just to set science up in business on the inhospitable ice."[14]

THE DEEP SEA OF ICE

Insofar as the astrophysical theory of synchronous ice ages skated on thin ice in Antarctica, it could be tested with a yardstick. Oil-industry geologists had developed one in the search for underground wealth. It was a technique of depth sounding, in which the echoes of shock waves set off by the explosion of dynamite or TNT could be used to identify and measure the depths of different kinds of rock.

This method of sounding the earth's surface, called explosion seismology, was first applied to ice in Antarctica by Thomas C. Poulter on the 1933–1935 Byrd expedition. Poulter, a geophysicist of the Stanford Research Institute of California, had been impressed by the acoustical properties of ice on the Ross Ice Shelf. It was such an excellent conductor of sound waves that it could carry ordinary conversation a hundred yards or more.

Poulter reasoned that it should be possible to measure ice thickness acoustically. Once the velocity of sound waves through a known depth of ice had been determined, the waves could be generated in ice of unknown depth and its thickness found by the time it took the waves to penetrate the ice and be reflected from rock bottom back to the surface, where they would be picked up and recorded by a seismograph.

After hundreds of experiments, Poulter was able to sound the thickness of the Ross Shelf, using small cans of explosives to generate the shock waves. Since the waves traveled through water at a velocity different from that through ice, he was able also to calculate the depth of the Ross Sea on which the shelf was floating.*

The technique of sounding ice by explosion seismology was further developed by the United States Army Corps of Engineers on the Greenland ice cap. It was not attempted on any major scale in Antarctica until 1949, when the Norwegian-British-Swedish expedition used it on a reconnaissance of Queen Maud Land. The party sounded the ice along a 375-mile route extending inland from the Weddell Sea. The soundings showed that the ice covered an Alpine landscape of buried mountains and deep gorges. In some places, it was 8200 feet deep.

When the International Geophysical Year opened in Antarctica, American research teams were equipped with such yardsticks, for a main objective in the Antarctic program of the United States National Committee of the IGY was to determine the thickness of the ice in West Antarctica and whether a great sub-sealevel trough lay underneath it between the Ross and Weddell Seas.

On the morning of January 28, 1957, five months before the IGY had begun officially, the first party of American investigators to probe these questions rolled out of Little America V on the Ross Shelf at Kainan Bay. So far as polar experience was concerned, the investigators were rank tenderfeet.

The team, in fact, consisted of a brand-new doctor of philosophy, three graduate students, and a mechanic, driving three gasoline-powered Tucker Sno-Cats. These vehicles, developed for the high sierras of California, were tractors with large cabs that looked like a cross between a small house trailer and a Sherman tank. The Sno-Cats had proved themselves rugged and durable during the construction phase of the scientific bases. Now they were being tried for the first time on a scientific traverse. The objective of the sortie was Byrd Station, 647 miles to the east on the flag-marked trail called the Army-Navy Highway. The party's mission was simply to get to Byrd Station, help complete its construction, and winter over there so as to get an early start, when spring came in October, to explore the mysterious region between the great embayments. En route, the scientists had been instructed to make seismic soundings as soon as they cleared

* Shock waves were reflected from both the ice–water interface and the water–sediment interface on the bottom. Ice and water depths could be determined separately also from bottom reflections. The ratio of ice thickness to water depth could be figured from comparisons of average wave velocities through both ice and water with theoretical velocities through columns of ice.

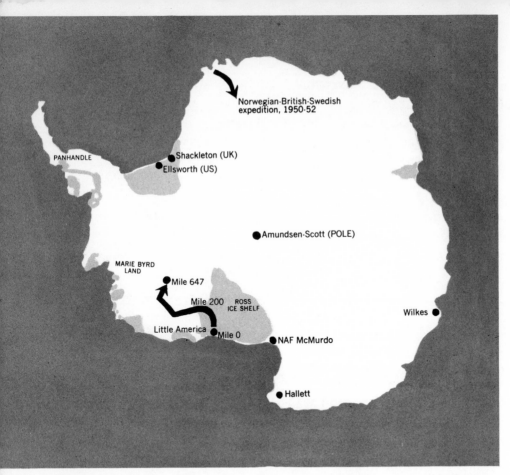

Bentley-Anderson traverse, Little America to Byrd Station, January 28 to February 27, 1957, five months before the official opening of IGY in Antarctica.

the ice shelf and found themselves on the grounded ice of Marie Byrd Land—about 200 trail miles from Little America.

This was, in effect, the first United States scientific traverse of the IGY in Antarctica. Its leaders were chief seismologist Charles R. Bentley, twenty-eight, who had received his doctorate from Columbia University a few days before leaving for Antarctica, and chief glaciologist Vernon H. Anderson, twenty-nine, a graduate student at the University of Wyoming. Ned A. Ostenso, a graduate student at the University of Wisconsin, was assistant seismologist, Mario B. Giovinetto of La Plata, Argentina, was assistant glaciologist, and Anthony J. Morency was the mechanic. All were on the scientific payroll of the Arctic Institute of North America.

Highway to Byrd Land

The caravan, its radio antenna waving in the cold breeze of late summer, moved out along the trail to Byrd Station. Each of the Sno-

Cats towed a 2.5-ton sled piled high with gasoline, food, and gear. Since each vehicle contained a cabin large enough for two to three men, it was unnecessary to pitch tents for sleeping or eating. The machines were powered by 8-cylinder engines and could travel up to 20 miles an hour over a smooth ice surface.

Buttons, the lead Sno-Cat, driven by Morency, flew an American flag. It was followed in single file by *Carole,* which carried the glaciologists and their equipment. *Hectori,* the third Sno-Cat, hauled the seismologists and their apparatus, bringing up the rear. It was named for *Gallirallus australis hectori,* a New Zealand bird.

For the first 7.5 miles the procession made grand progress. Then the radio, when tested, was found to be out of whack. The Sno-Cats were halted. After hours of tinkering with the radio, the explorers located the trouble in a defective vibrator. When it was replaced from the spare-parts kit, they were able to raise Byrd Station and explain their delay, to the amusement of the station personnel.

After this ignominious first day's run, the party had better luck. On January 29 the weather turned clear and sunny, with no wind. The temperature was a mild 15 degrees above zero Fahrenheit. *Buttons, Carole,* and *Hectori* performed nobly until they reached Mile 50 on the Army-Navy Highway. At this point *Carole's* fuel pump was leaking, and a halt was called to fix it. But finally the expedition was again on its way.

Sno-Cats poised to take off for a traverse led by Dr. Charles R. Bentley.

On the third day the party made 50 miles. Then *Carole*'s clutch began to slip. Time out for clutch repairs. These vehicles had been driven hard in the construction phase of Little America V. There had been no time to give them proper maintenance in the field. The traversers found themselves doing this on the trail, for in the Antarctic environment machinery is singularly unforgiving of neglect.

On February 1 the caravan passed through Fashion Lane, a treacherously crevassed area at the junction of the shelf and the grounded ice of Marie Byrd Land. Bridges of hardened snow, which concealed deep crevasses, had been dynamited away by the military party, and the crevasses—blue canyons sometimes 175 feet deep—had been filled in. Some of the wider chasms could not be filled, and the trail ran down into them and up the other side.

February begins the end of the austral summer. By the last week of the month, the sun is low in the sky. The temperature drops rapidly, far below zero, and the winds rise. Great storms of blowing snow, like desert sandstorms, sweep across the ice.

On February 2 the party reached Mile 200 on the trail, where its work of sounding the ice was to begin. Here was the "shore" of West Antarctica, the grounded ice stretching some 1800 miles across the cone of the Antarctic bell to the Weddell Sea and the Atlantic Ocean.

Up to this point the traversers had not been required to do anything but adjust to Antarctic travel. None of them had been in the Antarctic before, so each had his own peace to make with the environment. Sounding of the ice shelf was scheduled to be done the following season by another team under the polar veteran Albert P. Crary, then a geophysicist with the Air Force Research Laboratories at Cambridge, Massachusetts. The Bentley-Anderson traverse was to make soundings only on the grounded ice.

The halt at Mile 200 lasted two days. To the east the ice sloped upward from the shelf to the horizon, as though in conformity with a rising continental shoreline. Amundsen and his Norse crew had seen this slope forty-six years before as they drove their dogs southward. Admiral Byrd and Lincoln Ellsworth had seen it too and had assumed the slope of ice betrayed the contour of the land beneath. And if this was so, the ice had to be thin.

But if the land was rising, how could there be a channel or trough between the two embayments? Perhaps, as the Australian geologist Sir Douglas Mawson had speculated in 1928, the rising ice slope to the east of the shelf did not indicate land contour, but, instead, rested on a basement below sea level.

Near a parked Sno-Cat, Seabees take an ice coring to determine the thickness of the sea ice in McMurdo Sound.

Mount Erebus overlooking the junction of the ice shelf with Ross Island.

Bentley and Ostenso struggled to get the seismic rig set up, while Anderson and Giovinetto dug a 31-foot-deep pit in which they made temperature readings and measured snow strata. From the pit a cross section of the history of the climate in recent years could be read. The temperature at the bottom was about 27 degrees below zero Fahrenheit, representing the mean annual temperature of air over a period of snow accumulation for about 25 to 30 years. Along the sides of the pit, the snow in cross section showed lighter and darker bands, like rings of a tree. The darker bands represented summer snow, which is darker because it is denser and contains more dust blown in from the tropics. Counting from dark band to dark band, one could measure annual precipitation over a period of years. In this region, the width of the bands showed that less than 2 feet of snow fell a year—the equivalent of 2 inches of rainfall. Meteorologically, the area was a desert.

At some distance from the parked Sno-Cats the seismologists spread out their geophones and cable, in a large L-shaped pattern on the ice. The geophones would pick up the shock waves generated by an explosion of a half-pound of TNT and reflected to the surface of the ice from the rock beneath. The explosive was detonated in a hole 10 inches deep.

Bentley and Ostenso had to conduct a series of trials to find the average velocity of shock waves through the ice sections. They did not begin to get good reflections until the second day at Mile 200. These showed that the ice was less than 500 feet thick—thinner than the shelf ice they had left. They had reached high ground.

On February 5, the party packed gear back into the tractors and chugged on to Mile 250, at the rate of 5 miles an hour. Every fifth mile there was a 10-to-15-minute stop to take magnetic and gravity readings, which were other ways of sounding the ice.

To make a magnetic sounding, it was first necessary to estimate mathematically the strength and inclination of the earth's magnetic field for the site of the sounding. The difference between this calculation and the actual strength as shown by a device called a proton-precession magnetometer, which Bentley had managed to borrow from Columbia University, could be accounted for by ice. Once these differences, or anomalies, were calibrated against ice depths determined by seismic soundings, it was no trick to get fairly good estimates of ice thickness by this method.

Gravity soundings were made on the same principle. A theoretical value for the strength of gravity in the area was compared with the value shown on a gravimeter, a device which measures gravitational force, just as a magnetometer measures magnetic force. Matching actual with theoretical values required precise determination of latitude (with a theodolite) and of elevation (with an altimeter). As the force of gravity is inversely proportional to the distance from the earth's center, if the gravimeter showed a lower value than the theoretical one, the difference indicated a certain thickness of ice. When the anomalies had been calibrated with seismic measurements, a constant check on ice thickness could be maintained by this method.

Bentley had trouble at first with the magnetometer, which behaved crazily when he bent over to work with the instrument. At length he discovered it was reacting to his steel-rimmed glasses. After he took them off, the magnetometer worked perfectly.

With the seismic, magnetic, and gravity yardsticks, the party had a triple-check system of sounding the ice. All the men had to do was make it work. There was no manual to refer to if things went wrong. They were writing the manual that year.

At Mile 250 Bentley and Ostenso again laid out their geophones on the ice to pick up the reflected shock wave and transmit it to the seismograph. Boom went the dynamite. The results seemed a bit wild. They repeated the soundings again and again. The results were the same: the scientists were standing on a snow surface 2000 feet above sea level, but the seismograph persisted in giving reflection values which showed that rock bottom was 1500 feet below sea level. That meant the ice was 3500 feet thick! And that wasn't in the book, either.

Perhaps they had simply struck some kind of hole or cleft in the bottom surface. As they moved on, the soundings showed that the land continued to sink farther and farther below sea level, while the ice continued to rise above it. The divergence between ice surface and land was spreading. They were unquestionably over a very deep basin. And as they moved eastward, the basin became deeper.

At Mile 350, where the ice was sounded on February 9, the soundings showed it to be 1 mile thick. Explosion seismology was blasting a number of long-cherished conceptions apart. Here was no thin ice cap, nor did it follow the contour of the land beneath.

The ice thickens

During this period of the second week of February 1957 the weather was sunny and the winds were light. The Sno-Cats trundled eastward in the constant glare of the sun on a sea of ice—a sea which mile by mile seemed to be getting deeper.

At Mile 400 the party had ascended to an elevation of 3280 feet, and the ice was 6560 feet deep.

The West Antarctic ice cap was turning out to be thicker than anyone had predicted. The scientists were aware that their findings would be scrutinized minutely not only at Little America or in Washington, but at all the IGY data centers. They checked and double checked their results, fighting to maintain a rigorous working schedule which allowed a minimum of sleep. As long as the weather held, and the equipment continued to function, the work could be done.

Even the radio held up. Now and then the men searched for sounds of the world beyond the ice cap. On the night of February 10 Bentley was elated to hear an Armed Forces Radio broadcast direct from his home town, Rochester, New York. It was only the award of a belt by the Hickok Belt people to Mickey Mantle as professional athlete of the year—but it was a voice from home.

The radio was turned on from 10 p.m. to midnight (the party was operating on United States Central Standard Time, which was the time at Byrd Station). At midnight there were five minutes of the latest news—to which the young men who were making scientific history listened intently. Yet, the goings on in the outside world seemed curiously remote at times. "I feel so far from the world here," Bentley confided to the diary he had been asked to keep, "that the news seems of little importance any more."

All the men kept personal diaries. Not only was the practice *de rigueur* in polar regions but psychologists felt that the regular habit of recording the day's events would help the men adjust to the new environment and serve as an organizing ritual.

A gasoline cache was found at Mile 450, where the caravan parked to refuel and make soundings. Here the seismic reflections showed the rock bottom at 7800 feet below the ice surface. Beyond this point the ice thinned to 4600 feet for about 30 miles and then deepened again.

On February 17, Bentley and Ostenso, riding in *Hectori*, began to feel ill. Suddenly realizing that they were breathing carbon monoxide, they halted the vehicle and struggled out of the cab. The wind had been blowing in the same direction and at the same speed as the Sno-Cat was moving, so that the men had been enveloped in the vehicle's exhaust. This hazard was to plague other oversnow traverses under similar wind conditions.

The two seismologists took the day off to recuperate. After supper a tractor train from Byrd Station, grinding westward toward Little America, arrived and stopped long enough to deliver some mail, which had been air-dropped at Byrd. Bentley was interested to read that the winter temperature in upstate New York had hit an all-time low of 55 degrees below zero. It was 70 degrees warmer here on the West Antarctic ice cap, where the temperature that day was 15 degrees above zero Fahrenheit.

When they reached the 500th mile out of Little America, the party stepped up the frequency of pit studies and soundings to every 20 miles instead of every 50. The wind kept blowing in the direction of travel, keeping the Sno-Cats in a cloud of their own exhausts. The men traveled with their heads out of the window, which Bentley reported was "inconvenient." The more frequent working stops were a relief.

As the caravan approached Mile 600, the ice depth increased to 8200 feet. Seismic, magnetic, and gravity measurements showed that the vehicles were moving over a deep basin, the bottom of which lay 3500 feet below sea level.

February was ending. One day the sky became alive with rainbows. Shifting bands of red, green, and blue ringed the horizon. The display was created by the refraction of the lowering sun's rays by ice crystals in low cirrus clouds. Within a few hours the wind changed. The sky became overcast and the temperature plunged to 28 degrees below zero Fahrenheit. Autumn was coming to this silent land. At Mile 620 the party made its final working stop before Byrd Station. The ice was 8500 feet thick.

On February 27 the caravan began the final run. Within sight of the antenna masts of the station, *Carole*'s transfer case broke down and the vehicle had to be abandoned until a repair crew could rescue it. The five men rode into Byrd Station aboard *Buttons* and *Hectori*.

With them they brought seismograph paper tracings which revealed thick ice in West Antarctica. It was one of the truly significant geophysical discoveries of the IGY, and one of the most important in twentieth-century polar exploration. It upset estimates that had stood for fifty years. The books which talked of a thin-ice cap were now obsolete.

Ultimately the discovery of this first United States traverse of the IGY in Antarctica would change the concept of Antarctic physiography. Here was no mere residue of some past glacierization, but an ice age, full-blown, in the modern world.

The long day of summer became the twilight of autumn. Black and cold came the austral winter of 1957. There was only the light of the moon and of the polar stars on the ice cap until dawn came in August.

The third dimension of the ice cap had been perceived in all its magnitude for the first time. And yet the discovery that would change much of the thinking about this region, and about the past and future of the earth itself, was little noted in the world at that time. When Crary at Little America radioed the results of the Marie Byrd Land soundings to the National Academy of Sciences in Washington, it was suggested that he ask Bentley and Ostenso to recalculate their results. Crary replied tersely that this wasn't necessary. As senior geophysicist in Antarctica, he had the last word.

On May 1, 1957, while the winds shrieked between the half-buried huts of Byrd Station in the sub-zero darkness, a group of scientists and politicians assembled in a hearing room of the United States House of Representatives Office Building on Capitol Hill. It was spring in Washington. The cherry blossoms lay like tinted snowflakes on the green lawns. Among those present were Dr. Berkner, vice-president of the Special International Committee of the IGY; the late Dr. Harry Wexler, chief of the United States Weather Bureau Research and chief scientist of the United States Antarctic Program; and Hugh Odishaw, executive director of the United States National Committee of the IGY.

Representative Albert Thomas of Texas and Representative Sidney Yates of Illinois expressed their interest in reports that the ice in the vicinity of Byrd Station was 1.5 miles thick.

Dr. BERKNER: The chairman has put his finger on one of the most interesting discoveries in the Antarctic.

REP. THOMAS: Dr. Wexler, you would have to say you are not sure yet whether that ice could be several thousand feet thick?

Dr. WEXLER: Oh, yes. There may be 10,000 feet of ice.

Dr. ODISHAW: In the Antarctic, yes.

REP. YATES: Do you know whether that is the limit of the ice? Do you know that it does not extend beyond 10,000 feet?

Dr. WEXLER: For all we know, there may be 14,000 feet of ice. Or even more.

The investigators on the ice were carrying out a highly structured program of environmental investigation which had been evolving since 1953, when the United States National Committee for the IGY had formed its Antarctic Committee. It included research programs in meteorology, in the aurora, in the physics of the ionosphere, and in the nature and dimensions of the ice cap, as well as in the topography of the land beneath the ice. For these last, a far-ranging glaciology program requiring about 10,000 miles of travel on the ice cap was carried out with remarkable exactness and constitutes the most extensive traverse experience of mankind in the most inhospitable region on the planet.

The main objective was to "gather information on the present volume of the ice, the topography of the ice surface and topography and structure of the land beneath the ice."[1] In addition, the investigation was designed to "ascertain the present status of the Antarctic ice sheet, whether it is gaining or losing in mass or volume and the manner in which this gain or loss is taking place," and to probe the history of the ice sheet, the trend of the ice cap, and how the sheet responded to changes in solar radiation and ocean temperatures. Basically, the scheme was drawn to determine the structure of West Antarctica, and it went considerably beyond that objective.

How was it possible for a band of young men without any previous Antarctic experience to crack in five weeks a major physiographic problem that had baffled investigators for more than fifty years? One answer is that they were equipped with the proper tools and knew how to use them. But it is not the whole answer.

The successful investigation of Antarctic problems depended on a level of scientific and technological development that simply did not exist before the middle of the twentieth century. Mawson had suggested in 1928 that the most logical way of determining the structure of West Antarctica was to devise some kind of means of sounding the ice, but the means were not then available. The "means," it should be observed, are more than instruments. They include logistics—the ability to support and transport investigators on the ice cap. And the logistics of modern Antarctic exploration are based on the airplane, the icebreaker ship, and the tractor.

Scott had attempted to use motorized vehicles. Aircraft had been introduced in the Antarctic in the 1920s. But these vehicles were not developed fully enough to provide dependable transport, especially under Antarctic conditions. To explore such a rigorous region mankind had to achieve a state of technological competence equal to the task. Until it was achieved, exploration would be limited by the endurance of men and dogs. As Mark Twain said, "When it's steamboat time, you steam." And not before.

When the sun rose in August 1957 to illuminate the ice, the scientists of twelve nations prepared to mount the most extensive program of scientific investigation and exploration in history. This was the season when the 14,000-foot ice column was to be found—when the doctrine of thin ice would be blown away by explosion seismology.

From Cape Adare to the Weddell Sea, and around the great bell of East Antarctica, investigators probed the ice, charted the rocks, monitored the radiation from space that spiraled down magnetic lines of force into the polar atmosphere, studied the weather, observed the life of the sea and the coasts, and made endless photographs.

New and magnificent discoveries were made. But all of them were overshadowed by a 183.6-pound satellite called *Sputnik* and the dawn of the space age. *Sputnik* was the harbinger of an even newer and more extensive frontier. It was the forerunner of a dream at least as ancient as that which had conjured up Antarctica in the mind of man millennia before he found it. Having plunged deeply into the last land frontier on the planet, men now stood on the threshold of the first frontier of space.

But in Antarctica new knowledge had been found.

Crossing the ice bell

The discovery of a thick ice cap in West Antarctica by a group of young American scientists, though exciting scientifically, commanded less attention in the United States than the crossing of Antarctica by Sir Vivian Fuchs, the British scientist, with the aid of Sir Edmund Hillary of New Zealand, the conqueror of Mount Everest.

In America the British Trans-Antarctic Expedition of 1957–1958 came to be billed as a race to the pole between Fuchs and Hillary. This twist to the story borrowed heavily on the theme of Scott versus Amundsen and originated in a wire-service bureau in New Zealand. The fact that, by early 1958, Americans were serving coffee and doughnuts at the south pole to anyone who had the means to get there did little to dim the fascination of the "race for the pole" theme to news editors. Everyone could grasp the struggle of man against

Modern logistics for Antarctic exploration are based on the icebreaker ship, such as the one seen at top left escorting a cargo ship to its mooring site; the tractor to haul supplies across the ice from the ships and air fields; and the airplane, such as the C-130 Hercules at left, which makes long-range flights to Antarctica and between the Antarctic stations.

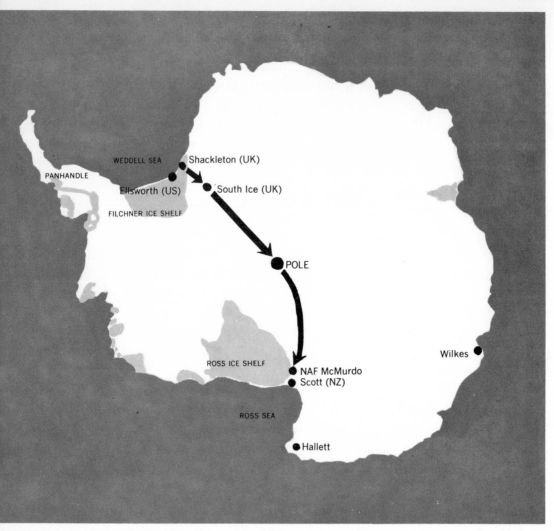

British crossing of Antarctica, November 24, 1957, to March 2, 1958.

man and of man against the elements in Antarctica. The race between Hillary and Fuchs was a harmless, if embarrassing, invention, but it fell into a gaslight frame of reference that news editors understood. The significance of finding deep ice in Antarctica seemed to have escaped them.

What Fuchs was trying to do was to bring to fruition Britain's old dream of crossing Antarctica, and Hillary had volunteered to help him. The deed had first been proposed in 1903 by the Scottish naturalist Bruce. Filchner had attempted it in 1912, and Shackleton in 1914. In fact, Shackleton was en route to Antarctica in 1922 for another go at it when he suffered a fatal heart attack.

The crossing of Antarctica between the Ross and Weddell Seas required bases at each end, well-stocked advance depots, and a mid-point supply cache. It was not until the IGY that such a 2200-mile journey became feasible, with Anglo-American air support and a well-equipped rest stop at the south pole.

Fuchs's Sno-Cats could make the run from the British Shackleton Station on the Filchner Ice Shelf to the pole with a cache of fuel at an advance inland base called South Ice. But the vehicles required additional fuel between the pole and McMurdo Sound. Hillary was charged with laying in fuel and food for the second leg of the trip.

Somehow, this logical and straightforward enterprise became altered into a highly romantic race for the pole. There were periodic reports of the progress of both teams, with Hillary "leading." Hillary reached Amundsen-Scott Station before Fuchs. When his arrival was reported by radio, news flashed around the world that Hillary had captured the south polar sweepstakes.

Fuchs was astonished by it all. His party picked up these reports on the radio as it made its tortuous way over the high polar plateau. In vain did Hillary protest that all he was trying to do was help "Bunny" Fuchs do a bit of science.

While the crossing of the ice cap was a magnificent piece of logistics, its scientific contribution has not been clear. The soundings by Fuchs's seismic men indicated a sub-ice mountain chain on the polar plateau, near Amundsen-Scott Station. When Fuchs's caravan began the 1250-mile trek to Ross Island along the 156th meridian, seismic soundings only a few miles from the station showed the ice to be less than 2000 feet thick. Fuchs concluded that "we had again reached a high rock area covered by a relatively thin mantle of ice. It therefore seems that the geographic South Pole is situated above a great ice-filled basin some 50 miles wide and lying between two rocky ranges that are themselves completely hidden by ice."[2]

American and Russian soundings do not show the sub-ice highlands. Instead, they depict the entire landscape beneath the polar plateau ice as an enormous basin, much more than 50 miles wide, depressed below sea level by the weight of nearly 2 miles of ice.

Assault on Victoria Land

In East Antarctica the United States program planned to make airborne soundings of the ice cap in Victoria Land in a preliminary survey of the region. After months of preparation, delays caused by lack of equipment, and other frustrations, two scientists flew to the Victoria Land plateau from the Naval Air Facility at McMurdo. They were John C. Cook, a seismologist of the Southwestern Research Institute at San Antonio, Texas, and William Vickers, a glaciologist of the Arctic Institute of North America. Luck, which is also an ingredient of Antarctic logistics, was not with them.

On January 2, 1958, the research men flew to the plateau in a ski-equipped Dakota R4D plane, the workhorse of Antarctic air transport during this period. The Navy pilot landed on the plateau and shut off the engines. In order to keep the engines from freezing, the aircraft crew attached specially designed heaters to them.

While the scientists spread their geophone net on the ice, one of the heaters caught fire and had to be ripped off the engine to save the aircraft. The unheated engine rapidly cooled. The scientists had difficulty interpreting the results of their reflections because of the rough, uneven rock bottom. But the threat of the unheated engine's freezing forced the party to leave before the soundings could be checked.

The plateau was obviously too high and cold for anyone to park an airplane there. On take-off the R4D's hydraulic system had become so sluggish that the controls could be operated only with great difficulty. On the next trip, it was decided, the airplane would leave the scientists for a day or two and return to pick them up.

A few days after the first sortie, the scientists were landed on the plateau with their equipment, tents, and enough food for three days, though the aircraft was to return for them in twenty-four hours.

Delayed by an errand at the Navy base, the plane did not show up at the appointed hour. In vain the scientists, who had worked fran-

Silhouette of a ski-equipped Dakota airplane in early October.

tically to complete their soundings, with only a five-hour break in twenty-four hours, searched the sky. All they saw were signs of a blowing snowstorm in the west. These storms, which are the Antarctic equivalent of desert sandstorms, sometimes rage for days.

At the end of six anxious hours the men heard the drone of the Dakota's engines in the vast silence. Once they had the plane in line of sight, they called it down on their transceiver radios. The pilot explained he had been flying a search pattern and was low on fuel. He was relieved that he had found the camp at all.

The seismic data on the two airborne excursions to the plateau indicated ice depths of 12,500 feet. But only three locations were represented in the survey. It was clear that Victoria Land soundings called for a fully equipped tractor expedition. Airplanes would do better on the lower and warmer slopes of West Antarctica.

It remained for Albert P. Crary, then deputy chief scientist of the Antarctic program, to make the first full-scale assessment of the Ross Ice Shelf and to undertake extensive soundings in Victoria Land.

Though the shelf, a pie-shaped wedge of ice as large as France or Spain, had been known for more than a century, its full dimensions had never been measured, nor had its connection with the continental ice cap been established. Crary's objective was to determine its area, its thickness, where it was afloat, where it was grounded, and its relationship to the main ice sheet.

The Ross Ice Shelf traverse set out from Little America Station on October 24, 1957, in three Sno-Cats. The party circumnavigated the shelf in 3½ months, covering 1450 miles, and returned to Little America February 13, 1958.

Crary determined that the shelf has an area of 210,000 square miles. Its thickness ranges from 790 feet on the Victoria Land side to 1050 feet on the Byrd side. The deepest ice was found where the shelf abuts the mountains of the polar plateau.

A portion of the shelf on the eastern side was apparently aground on a large island called Roosevelt Island, but most of it floated on seawater 2070 to 2155 feet deep. The measurements were of significance in relating the shelf to the grounded ice cap.

The shelf seemed to represent an extrusion of the continental ice sheet over the Ross Sea. If the source of its nourishment could be found, it would indicate in which direction a vast part of the continental ice sheet was moving. The direction and rate of flow of the ice cap are essential clues in determining whether the ice cap is waxing or waning.

It was obvious that the shelf was receiving ice from the polar plateau. Great rivers of ice poured through the mountain passes, the largest being the Beardmore Glacier. As he expected, Crary found that the shelf was thicker at the Beardmore (1378 feet) than at Little America Station (951 feet) on its northern rim. Also, it was thicker on the Byrd Land side than on the Victoria Land side, indicating that the shelf in part is an extension of the West Antarctic cap and of the polar plateau ice, rather than of the sheet in Victoria Land. Since the shelf was thinner on the Victoria Land side, it apparently was not getting much ice from that segment of the Antarctic ice sheet.

This finding suggested an interesting point about the East Antarctic sheet in Victoria Land. The ice beyond those shining mountains must be moving away from the shelf, toward the Pacific Ocean. Crary wanted to drive the Sno-Cats up onto the Victoria plateau and attempt to confirm this idea. But by the time he had rounded the shelf the season was too far gone to extend the traverse.

"More ice than anyone thought . . ."

The next austral summer, 1958–1959, Crary led a second expedition out of Little America, crossed the shelf to McMurdo Sound, and ascended the Skelton Glacier to the high Victoria plateau. Beyond the mountains, Crary and his colleagues found that the ice surface fell 500 feet in elevation. It certainly appeared from this that the plateau ice was flowing not into the Ross Shelf but the other way, toward the Pacific, as Crary had speculated.

As the expedition probed westward the elevation of the ice surface increased from 6000 to 9348 feet. At this altitude in polar regions breathing is more difficult because of lower oxygen pressure than at the same elevation in temperate or tropical zones, an effect of slightly higher gravity and low temperature.* Seismic reflections showed that the land lay at sea level, or slightly below. If relieved of the great load of ice it bore, the land might be on the order of several hundred feet above sea level. The high plateau of Victoria Land, like the polar plateau, was ice, all ice—a giant cake of ice nearly two miles thick, with an area of 1,000,000 square miles or more.

* The force of gravity at the poles is 0.5 per cent higher than at the equator, which means that a 200-pound man, for example, gains a pound in weight by traveling from the tropics to the pole. This increase of weight is a result of two factors: because of the oblateness of the earth's surface, the poles are closer to the center of the earth, and the counter pull against the gravitational attraction, caused by the earth's rotation, which is greatest at the equator, is nil at the poles.

Crary, given to understatement, commented that "there was a lot more ice on the plateau than anyone had thought." There was a lot more in Antarctica than anyone had guessed—nine-tenths of all the ice in the world.

Crary had hoped to reach the Soviet Union station, Vostok (East), about 1000 miles inland. But the Navy's VX-6 Squadron based at NAF McMurdo could not supply the expedition at that distance. After driving 700 miles into the Victoria Land interior, Crary and his party turned their Sno-Cats back east and returned to McMurdo.

Members of the 1960 traverse to the south pole, led by Albert P. Crary (standing second from the left).

A Rolligon, loaded with equipment for the Crary traverse, is towed by a Diesel-powered Sno-Cat.

At Vostok, Russian seismologists reported ice depths of 11,217 feet and an average thickness in other areas of 8700 feet. The French, ranging inland from Charcot Station in Wilkes Land on the Pacific Coast, found the ice generally 9000 feet thick. At the south pole American and Russian soundings showed that rock bottom lay 9200 feet below the surface of the snow.

The following summer, 1959–1960, another American expedition tackled the plateau. It was led by the Dutch geophysicist Franz Gustaaf Van der Hoeven, Jr., a member of the staff of the Lamont Geological Observatory of Columbia University. The party included New Zealand scientists and a psychologist from Georgetown University, who went along to see how men react under varying conditions of an oversnow traverse and became the expedition's cook. This expedition covered a rectangular route on the plateau of 1500 miles. Its soundings showed ice depths ranging from 8000 to 12,000 feet.

In the summer of 1960–1961 Crary led a Sno-Cat expedition from McMurdo Sound to the south pole via the Skelton Glacier and the Victoria plateau. The party used new and more powerful Sno-Cats with Diesel engines. Each of the tracked vehicles towed a cargo platform called a Rolligon, slung between four huge rubber tires and capable of carrying two to four tons of equipment. A unique feature of this rig is that each of the tires can be filled with 500 gallons of Diesel oil instead of with air. When needed on the traverse, the oil can be drained and replaced with air to keep the tires inflated.

En route, soundings made by the party on the polar plateau failed to find evidence of the sub-ice mountains reported by Fuchs. After reaching the pole, the party, which included a Soviet exchange glaciologist, flew back to McMurdo Sound, leaving the Sno-Cats and Rolligons for further use at the pole station.

This expedition capped four years of oversnow investigation by the United States and other countries, in which the thickness of the Antarctic ice sheet was established. American expeditions had sampled about two-fifths of the sheet. The results and those of other nations had yielded one of the most important geophysical discoveries of modern times: the Antarctic ice cap was a thick one, with an average depth of 8000 feet and an estimated volume of 8,000,000 cubic miles. The water locked up in it was of the order of the North Atlantic Ocean.

The assessment of the ice cap meant that all estimates of the earth's water budget up to 1958 had to be revised upward by at least 20 per

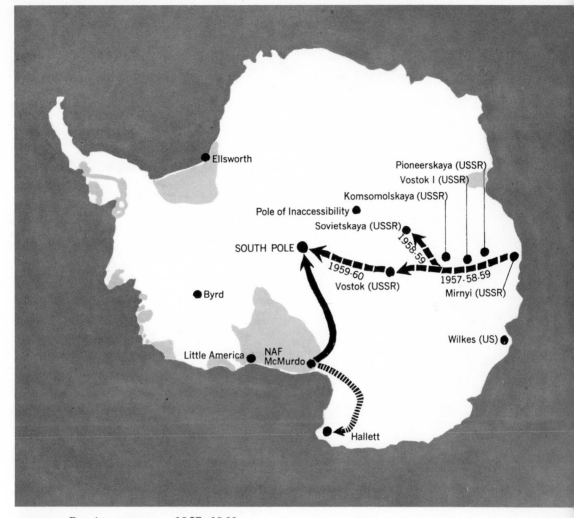

Map labels: Ellsworth, Pioneerskaya (USSR), Vostok I (USSR), Komsomolskaya (USSR), Pole of Inaccessibility, Sovietskaya (USSR), 1958-59, SOUTH POLE, 1959-60, Vostok (USSR), 1957-58-59, Mirnyi (USSR), Byrd, Wilkes (US), Little America, NAF McMurdo, Hallett

■ ■ ■ ■ *Russian traverses, 1957–1960.*
■■■■■■■ *Crary's traverse, McMurdo Sound to the Pole, 1960–1961.*
ıllıllllllllllllllı *Van der Hoeven's Victoria Land traverse from McMurdo Sound, 1959–1960.*

cent. Now for the first time it was possible to make an inventory of the planet's full water resources.

The investigation showed that in Antarctica there was an ice age in full swing—not the residue of one past. If ice ages were synchronous in the hemispheres, how could this be? Yet there had to be simultaneous glacierizations in the north and south if one accounted for climatic changes by astrophysical theory. These findings in Antarctica implied that changes in solar energy were not the whole story. There might be another answer.

RIDDLE OF THE LAND

BENEATH THE ICE

The structure of the land beneath the ice of Antarctica represented one of the major physiographic and geographic problems of the twentieth century, inherited from the nineteenth. It may be that the twenty-first century will continue to struggle with it, for it is not solved yet.

Determining the nature of the ancient Antarctic landscape is a project comparable to that of exploring the whole of the United States and of Mexico if they were covered by 7500 feet of ice. Obviously the exploration of 5,500,000 square miles of territory under such conditions could not be accomplished in a few scattered expeditions in which men expended half their energy battling the most hostile environment on earth. Yet most of what we knew about Antarctica up to 1957 was based both on the findings and guesswork of intermittent sorties over the southern ice cap. It is no wonder that the general conception of these lands was ambiguous.

On a spherical map of the earth—a terrestrial globe—Antarctica conventionally is outlined in white at the bottom as a bell-shaped figure, taken to represent the Antarctic continent. But what we see on the map is not a continent; it is an ice cap. And the two are not the same. The supposition that they were has been the biggest error in geography since Columbus believed he had reached India.

Allied to the notion that the ice and land are the same was the fallacious idea that the contour of the ice surface followed that of the land beneath. This resulted in the assumption that the ice was thin, and, in turn, that idea formed one of the pillars of the astrophysical theory of synchronous ice ages.

All these erroneous ideas were exploded by seismic, gravity, and magnetic soundings of the first American expedition to cross the West Antarctic ice sheet in the IGY. It was not until then that we began to perceive the diversity between the ice surface and the underlying landscape.

Today we still know less about Antarctic topography than about the surface of the moon. The lunar hemisphere facing earth at least can be seen, but 98 per cent of Antarctic land is concealed by ice. Even the mapping of the visible land—the mountains protruding through the snow—has been subject to error. Mountain peaks charted by one expedition could not be found when the sites were visited by another. And mountains keep popping up in Antarctica where the map says no mountains ought to be.

It is reasonable to ask: why bother about a land surface man may never see? But until he knows the geology of the land the scientist cannot be certain about the relationship of Antarctica and the other lands of the southern hemisphere. Nor can he account for the geophysical history of the earth and the events which brought about the great ice cap. Antarctica is a key to the evolution of the earth. Without it, the mystery of the planet's past may never be unlocked, and its future may never be predicted.

This riddle—the nature of the land beneath the ice—has been the hardest of all. Out of six years of soundings, a rule emerged that the more Antarctic regions are investigated the more complex they become.

It had all seemed so beautifully simple and clear at the end of the nineteenth century, when evidence that a lost continent did indeed exist beneath the ice was offered by Murray. Such a view, without complication, was possible only from a distance. Closer inspection by Borchgrevink, Scott, Shackleton, Taylor, Priestley, Mawson, and others revealed a mysterious dualism in the buried landscape. West Antarctica and East Antarctica were not part of the same pie. Griffith Taylor and Raymond Priestley had raised the question of whether there was one continent here or two, and Taylor's speculation that these two geotectonic provinces were divided by a deep trough between the Ross and Weddell Seas gave form to the problem.

By 1910, Murray's Hypothetical Continent had proliferated into possibly two hypothetical continents, with a hypothetical trough or arm of the sea running between them. For the next forty-seven years evidence for and against the trough accumulated. Amundsen had seen the rising surface of what is now called Marie Byrd Land to the east of his track as he raced across the Ross Ice Shelf toward the polar mountains. That was taken as an indication of rising land, precluding a troughlike depression or sunken valley. On the other hand, Filchner made soundings in the Weddell Sea in 1912 that showed there was a deep trough at sea bottom which extended under the ice shelf in the direction of Byrd Land. Shackleton repeated the soundings in 1914 and got the same results.

Mawson, impressed with Taylor's argument for the trough, said, "It may be that the rising ice slopes seen by Amundsen were resting on a basement still below sea level. In this case, were the Antarctic ice to melt and disappear, a small sea channel might connect the Pacific and Atlantic oceans."[1]

Byrd did not agree. His reconnaissance aircraft had sighted an ice plateau at least 1000 feet above sea level. The region was "land, all land," he wrote. "Were the ice to melt it would stand as solidly interposed to the questing keel of a ship as the Atlantic Coast stood before the sixteenth-century navigators seeking a strait to Cathay."[2]

The contour of the land remained unsettled so far as the West Antarctic region was concerned. In 1928 a leading British geographer, R. N. Rudmose Brown, wrote in a quarter-century survey of geophysical problems, "The broad features of the map of Antarctica are not built on ascertained fact as much as on intelligent guesswork. The existence of an Antarctic continent is still based on circumstantial evidence. It is not a little remarkable that all the exploration of the twentieth century has merely modified the probable outline of the continent as it was predicted."[3]

Byrd added in 1935: "It's a curious but readily accredited fact that long after most astronomers had settled to their satisfaction that there were no canals on Mars no geographer nor geologist could have told you whether Antarctica only 10,000 miles away was one continent or two."[4]

A peripatetic penguin

The initial traverse of Bentley and Anderson from Little America to Byrd Station had not only disproved the thin-ice notion but also the idea that the ice followed the land contour. Nothing could have been further from the strange reality of Marie Byrd Land. For in spite of the rising ice surface observed by Amundsen and Byrd, the traverse had sounded a huge basin lying below sea level, just as Taylor had predicted.

In the waning summer of 1957 Anderson and Bentley had followed the trough only as far as Byrd Station. To the east the icy waste had been reconnoitered by Byrd and Lincoln Ellsworth from the air. Ellsworth had mapped a lonely mountain range whose peaks stood up in the snow like sentinels, guarding the east, and had named the range the Sentinel Mountains.

Did the trough extend from Byrd to the Sentinels? Were they a barrier to its continuation to the Weddell Sea? These questions were foremost as Bentley and Anderson led their second traverse into the vastness of West Antarctica. They left Byrd Station on November 19, 1957, to sound the ice at least as far as the Sentinels.

Ostenso again was assistant seismologist. Daniel P. Hale, an aurora observer at Byrd, joined the trek as physicist. William E. Long, a graduate student at Ohio State University, was assistant glaciologist. His brother, Jack, of Richmond, California, was the mechanic.

The party drove the trusty tractors *Buttons, Carole,* and *Hectori* northward toward the coast on the first leg of the journey. The ice continued to sound at depths of 8200 to 8500 feet, with rock bottom 3280 feet below sea level.

By the end of November, the men sighted the peaks of the volcanic ranges of the Marie Byrd Land coast. As they approached these conical peaks, the trough became shallower. They had reached its northern "shore." The land rose toward sea level at a point about 250 miles north of the station. This leg of the traverse had shown the width of the trough.

The party turned eastward, along the trough's major axis, where the ice thickened. Depths increased to 9512 and 10,824 feet. Much later in the traverse the soundings reported depths of 11,611 feet and 13,940 feet—the deepest ice found in the world up to that time.

What were the Sentinel Mountains? Were they an extension of the Andean structure on the Panhandle? Did they block the trough? Did they butt up against the Trans-Antarctics on the polar plateau? These were the principal questions looming ahead of the expedition as it moved slowly eastward.

During the second week of December, the traversers reached a single peak jutting out of the snow. These rocks, which rise from a few to several hundred feet above the snow, are called "nunataks,"

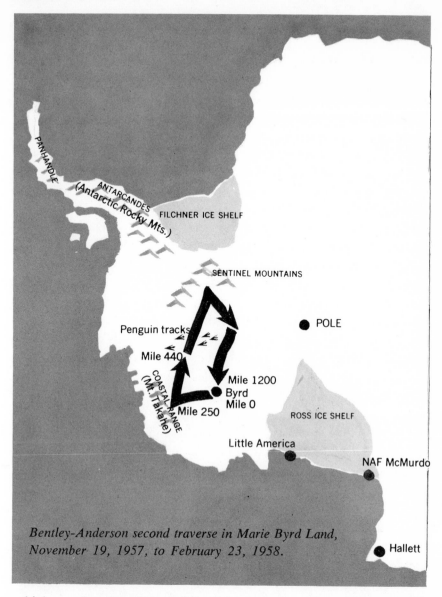

PANHANDLE

ANTARCANDES
(Antarctic Rocky Mts.)

FILCHNER ICE SHELF

SENTINEL MOUNTAINS

Penguin tracks

POLE

Mile 440

COASTAL RANGE
(Mt. Takahe)

Mile 1200
Byrd
Mile 0

Mile 250

ROSS ICE SHELF

Little America

NAF McMurdo

Hallett

Bentley-Anderson second traverse in Marie Byrd Land,
November 19, 1957, to February 23, 1958.

which are the pinnacles of drowned mountains. The party attempted to drive the Sno-Cats up onto the bare rock, but should have known better. A full half-mile from the nunatak the rear pontoon of *Carole* slid into a crevasse which had been obscured by a light bridge of snow. It took all hands four hours to lift the vehicle out. The three Sno-Cats were then backed off and the party hiked to the nunatak to collect rock samples.

The nunatak was a burned-out volcanic cone, probably a part of a mountain group called the Crary Range. It was agreed to name the cone Mount Takahe—after a supposedly extinct New Zealand bird—so that the cone could serve as a reference point.

From Mount Takahe the traverse turned to the east on a beeline for the Sentinels. The ice became thicker but the surface elevation remained the same. It was clear that the party was once more over the trough.

As the vehicles rolled on, the trough deepened. It fell to 4592 feet below sea level. On Christmas Day, the soundings showed its depth was 6500 feet. The resupply airplane from Byrd Station arrived Christmas afternoon with two chaplains to conduct Protestant and Catholic services under the open sky. When the services were concluded, the Navy R4D aircraft whooshed off the ice on a pillar of fire and smoke from its JATO bottles. The investigators climbed aboard their vehicles and rolled on into the wilderness.

The trough continued to the east. There was little doubt in Bentley's mind that it was sea bottom, choked with ice. With the ice removed, it would be a sea, possibly 400 miles wide from north to south. By New Year's Day, 1958, the scientists had traced its length for 700 miles from the Ross Ice Shelf. The day was celebrated with a dinner of fried chicken and strawberry ice cream.

On January 2, at a point 440 miles from Byrd Station, a curious discovery was made. There in the trackless waste was a track—a penguin track that appeared out of nowhere—at least 186 miles from the nearest known coast. And, curiously enough, the track showed that the pudgy little bird was headed east—away from the coast. Penguins had been known to wander 30 to 40 miles from their coastal rookeries, but no evidence of them had ever been found so far inland.

"We followed the track for a mile or so," Bentley recorded in his diary. "Its direction never varied. We collected two samples of excrement, measured the size of the belly-slide marks and the foot marks and found one place where he had stood up for a few steps. No doubt about it being a penguin. Where on earth could he have been going? How did he keep going so straight, especially in an area of rough sastrugi [surface wavelike ridges in the ice] which were over his head? The whole thing is fantastic, yet the track was really unmistakable."

The penguin had walked for only two yards. The rest of the way, the track revealed, it had tobogganed on its belly—a standard locomotion procedure for penguins, which can propel themselves quite rapidly over the ice, using their flippers and feet.

Photographs of the track made by Jack Long were examined months later by Dr. William J. L. Sladen, a British research biologist at Johns Hopkins University. He concluded that the track was made by an Adélie penguin and that the straight-line direction of it indicated that the bird was on a purposeful journey, not simply wandering.

The implications of the track were interesting. Penguins, like most other birds, have a built-in homing mechanism. It enables them to return to their rookeries or food sources from truly great distances—in one substantiated case as much as 2400 miles. Since these birds don't fly, that is quite a feat. It would not have been possible for this bird, no taller than a one-year-old child, to have climbed to an elevation of 4800 feet from the coast unless there had been an easy approach to the interior ice cap. The nearest coastline was the unexplored Eights Coast, of the Bellingshausen Sea. Hence there was probably some kind of an embayment, from which the bird could get sea food, indenting the ice in this region. Sladen suggested this was easier to believe than that the bird had traveled nearly 200 miles on a barren, foodless ice cap.

This line of reasoning suggested that the trough, or a branch of it, extended to the Bellingshausen Sea. Since this sea is east of the Antarctic Peninsula, such a trend in the trough would indicate that the Peninsula was not connected to the West Antarctic bell, from which it seems to extend as a handle. Early speculation had it that the trough ran to the Weddell Sea, west of the Panhandle, making the handle of the bell a part of West Antarctica. The penguin track

The Adélie penguin uses its strong flippers for swimming and propelling itself across the snow and ice.

implied a different structural relationship of the two Antarcticas from what had been supposed for fifty years. Later we shall see how important this clue turned out to be.

The bird that left the clue was never found. The expedition moved on. On January 7, 1958, the Sentinels were sighted. They rose up above the ice plain in a procession of ten peaks ranging north and south. As the scientists drew nearer, the mountains took on a scalloped, clamshell contour against the blue-green sky. At 400 miles east of Mount Takahe the soundings revealed that the trough was coming to an end; the rock bottom was rising to sea level.

From the Ross Shelf the scientific team had followed the channel for 1100 miles. Here a mountain wall crossed the axis of the trough like a dam. Now the question was: what became of the West Antarctic channel at the Sentinel Mountains? Where—if indeed anywhere—did it go from there?

Within thirty miles of the range the party was delayed for two days by a blinding snowstorm. Then, on January 11, as the weather cleared, the men drove the Sno-Cats nearly to the west slopes. The expedition halted. Between it and the main rocks lay a crevassed valley six miles wide—impassable. Turning south, the men drove along the edge of the treacherous region, looking for a way through. The moat was unbroken, guarding the rocky battlements from any approach over the ice.

"What a range it is," Bentley noted in his diary, "full of extremely steep faces and jagged edges. Appears to be tightly folded sediments of several kinds, judging from the color chiefly."

Comparison of magnetic values with rock topography, from data of the Byrd Station traverse, 1957–1958.

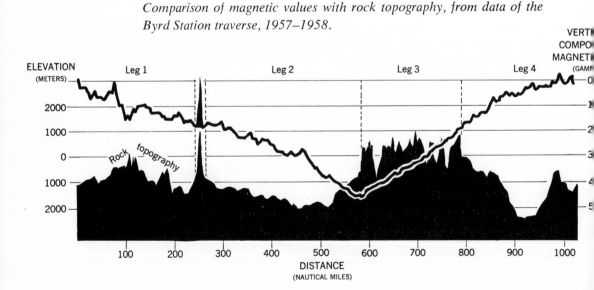

The scientists were able to reach three outlying peaks and collected samples of low-grade metamorphic rocks. Bill and Jack Long and Ostenso climbed one peak and found on the summit, 1000 feet above the snow surface, glacial striations—evidence that the ice had been thicker in the past. But whether the recession was local or general, whether it represented a climatic change or simply a decrease in snowfall, these were questions that only a full-scale assessment of the ice cap could answer.

The structure of visibly folded metamorphic rock proclaimed the affinity of the Sentinels to the Antarctandes of the Panhandle. Biological clues gathered by Hale, the party's physicist, seemed to offer confirmation.

Hale collected sixty specimens of lichens on the exposed rocks. These hardy, slow-growing plants, a symbiotic combination of algae and fungi, are the only vegetation on the ice cap. He found that individual peaks, though fairly close together, had different species of lichens, most of which could be linked to species previously found in the Panhandle. It seemed reasonable to assume that the lichens had propagated southward from the peninsular mountains.

Insofar as it could be gathered the geologic evidence showed that the Sentinels represented an extension of the Andean system into the heart of West Antarctica. By itself, this was an illuminating discovery, implying unforeseen geographical and even political relationships with the Americas that will be discussed later.

The party followed the chain southward for 186 miles. Except for two narrow gorges the land remained above sea level. In the south the Sentinels sank below the surface of ice. Isolated nunataks extending far into the south toward the polar plateau suggested that the Andean extension continued to the Pre-Cambrian Trans-Antarctic Mountains. If this was the case, there could not be a channel connecting the Ross and Weddell Seas.

Taylor's hypothetical trough, so brilliantly demonstrated at the outset of the investigation, had vanished in the Sentinel Mountains. Or so it seemed as the lowering sun and the rising winds forced the expedition to head back to Byrd Station in February 1958.

The Grand Chasm

Nothing is definitive in Antarctica. No evidence is final. No speculation is safe. No one can outguess this region.

The great trough which had vanished at the Sentinels in the region of West Antarctica called Ellsworth Land reappeared under the

Filchner Ice Shelf in the Weddell Sea. It was traced inland by the Ellsworth Station expedition in a survey of the Filchner Ice Shelf. But this expedition did not have time to pursue it into Ellsworth Land —if indeed it went there.

The Ellsworth Traverse circumnavigated the Filchner Ice Shelf for the first time and discovered this great mass of floating ice was more than four times as large as had been supposed. It was, in fact, second in extent only to the Ross Ice Shelf and had an area two-thirds that of France. The leaders of the traverse were the late Edward C. Thiel, then assistant professor of geology at the University of Wisconsin,* and Hugo A. C. Neuburg, then a graduate student at New York University.

Unlike the Ross Shelf the Filchner had been barely explored except for aerial surveys in 1947–1948 by Finn Ronne. Ten years later Ellsworth Station had been erected on the shelf as a base for the exploration of the Weddell Sea area. Breasting the Weddell Sea to reach the shelf had been no picnic. The Navy cargo ship *Wyandot* and the icebreaker *Staten Island* spent 43 days in the ice pack on a sail of more than 2000 miles during the 1956–1957 austral summer, searching for a site.

The plan was to erect Ellsworth Station on the Panhandle east coast, but ice conditions in the Weddell Sea that year were so bad the ships could not get near the coast. They were constantly threatened with being beset in the ice, like the ships of Filchner and Amundsen in earlier times. Hence the Navy had to compromise on a station site on the ice shelf itself. The ships unloaded 6000 tons of supplies, 9 scientists, and 30 Seabees of the Atlantic Fleet, and then sailed away to escape being frozen in. All hands settled down for the winter, after putting up the station buildings under the direction of Finn Ronne, the station commander.

When spring came, in October 1957, Thiel and Neuburg found that the clutch springs of the Sno-Cats to be used in exploring the ice shelf were shot. The vehicles had been overworked during the construction of the station, and there were no spare vehicles or parts.

Neuburg flew up the coast to the neighboring British Shackleton Station, hoping to borrow springs from Fuchs, whose party was preparing for the historic crossing of Antarctica. Sorry, old chap, he was told; Fuchs had no spares either. Having borrowed a set of springs as a model, Neuburg flew back to Ellsworth. He painstakingly retemp-

---

* Dr. Thiel was killed in an airplane crash at Wilkes Station in 1962. He was then on the faculty of the University of Minnesota.

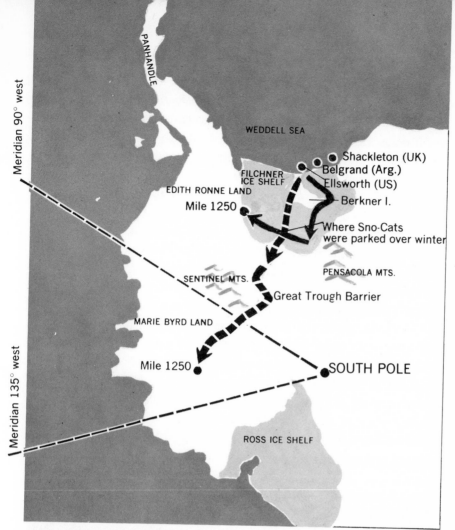

Meridian 90° west

PANHANDLE

WEDDELL SEA

Shackleton (UK)
Belgrand (Arg.)
Ellsworth (US)
Berkner I.

FILCHNER
ICE SHELF
EDITH RONNE LAND

Mile 1250

Where Sno-Cats
were parked over winter

SENTINEL MTS.

PENSACOLA MTS.

Great Trough Barrier

MARIE BYRD LAND

Mile 1250

SOUTH POLE

Meridian 135° west

ROSS ICE SHELF

▬▬▬ *Filchner Ice Shelf traverse of Thiel and Neuburg, October 28, 1957,
to January 17, 1958.*

▪▪▪▪ *Ellsworth-Byrd traverse in Edith Ronne Land and Ellsworth Highland, October 31, 1958, to January 15, 1959.*

ered the burned-out springs over the kitchen stove. Then he returned
Fuchs's clutch springs to Shackleton Station. Both expeditions set off
about the same time.

Thanks to the British Trans-Antarctic Expedition, Thiel, Neuburg,
and their associates were able to change the map of Antarctica. Up
to that time the Filchner Ice Shelf was shown on all Antarctic maps
as only a thin rind of ice. The Thiel-Neuburg traverse found, instead,
a shelf ice area of 140,000 square miles, split by what appeared
to be a large island. It was another great feature of the Antarctic
icescape whose magnitude had been underestimated.

During the construction of Ellsworth Station early in 1957, there
was time for several airplane reconnaissance flights over the shelf

before the sun set in March for the winter. On one flight an amazing feature of the shelf was discovered—a rift in the ice about 50 miles south of the station. The split, which ran east to west for a distance of 60 miles, ranged in width from 0.25 mile to 3 miles. Its apparent depth was 175 feet, although it might have extended all the way through the shelf.

The huge crevasse, which became known as the Grand Chasm, effectively fenced in the Ellsworth Station from the interior so that any expedition heading south would have to skirt it. This had to be taken into account in the traverse planned by Thiel and Neuburg to investigate the interior of the Filchner Ice Shelf, which was still unknown.

The Filchner Traverse party drove out of the station on October 28, 1957, in two Sno-Cats, each towing a sled with two and one-half tons of supplies. In addition to Thiel and Neuburg, members of the party were Paul T. Walker of Ohio State University, John C. Behrendt of the University of Wisconsin, and Nolan B. Aughenbaugh of Purdue University. While a DeHavilland "Otter" airplane scouted ahead, the traversers made their way slowly along the north rim of the Grand Chasm. When they reached the eastern end of the cleft, they began nosing the Sno-Cats through the crevasse fields which bordered it, guided by radioed instructions from the air.

After three tense days of negotiating crevasses, the party emerged on safe ice and turned south. Almost at once seismic soundings by Thiel and Behrendt picked up a deep trough beneath the shelf. Thiel had expected it since the downwarping had been recorded at sea on the fathometer of the *Wyandot*. The expedition followed the trough to a point 160 miles south of the ice front, where it passed between a big snow-covered island in the shelf and the Pensacola Mountains. In most places the trough sounded at 3500 feet below

Profile showing the depth of the land beneath the water and ice, from soundings taken in the Filchner Ice Shelf traverse, 1957–1958.

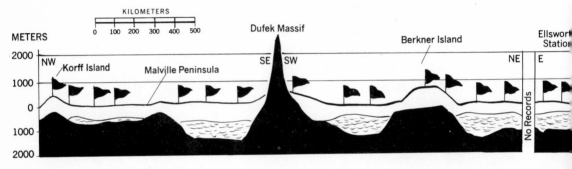

sea level—about 2000 to 2500 feet below the Weddell Sea bottom. It appeared that the sub-ice channel continued toward the Ellsworth Highland, and it was tempting to surmise that it was the Atlantic end of a Trans-Antarctic trough. Thiel and Neuburg were not aware that Anderson and Bentley had found the trough cut off at the Sentinel Mountains.

The two expeditions, out on the ice at the same time, had planned to confer by radio, but neither was able to make contact with the other because of intermittent radio blackouts. For several weeks the Ellsworth Traverse could not even raise its own station.

The Filchner Traverse turned aside from the trough and ascended the ice-covered central island. Thiel later concluded that the island is actually a peninsula, joined to the mainland, and dividing the ice shelf into two unequal parts. The highland area rose to 2500 feet above sea level, and the scientists estimated it was about the size of the state of Delaware. Twice the Sno-Cats sank into shallow crevasses as the party worked its way up a steep slope to cross the southeastern corner of the highland. The elevated region was named Berkner Island, in honor of Lloyd V. Berkner, the chairman of the United States National Committee for the IGY.

Leaving the so-called island, the traverse party turned south toward a range of mountains which had been sighted first in 1956 on a Navy reconnaissance flight from the Ross Sea to the Weddell Sea, commanded by William M. (Trigger) Hawkes, who dubbed the range the Pensacola Mountains, after the Naval Air station in Florida. On the northern end of the range, a great block of rock named the Dufek Massif, after Admiral Dufek, reared up out of the snow.

Four hundred miles south of the ice front the scientists reached the Dufek Massif, where the shelf ended. In the rocks they found abundant green stains of malachite (copper carbonate) and mineralized bands of iron and chromium—clues to the mineral wealth buried in the continental shield of East Antarctica under thousands of feet of ice. But as far as these minerals were concerned, the isolation of the range "prohibits any economic importance at the present time," the scientists reported.

As in the Sentinels, there was evidence of a once higher ice cap. The lower end of the massif had been rounded off, indicating that ice had overridden it in the past, and moraines, the telltale piles of rock swept up by advancing glaciers, were found in parts of the massif from which the ice had retreated.

Deeper in the mountains the investigators came upon dry, ice-free valleys, with beds of crushed rock. In one valley was a melt-water lake, 300 yards wide, its bottom covered by a pink plant, about 4 inches high, later identified as an alga called *Phormidium incrustatum*.

Along much of the northern slope of the massif the ice had vanished, leaving patches of bare, rock-strewn earth as large as 10 miles square. Such a sight is rare in Antarctica. Eleven thousand years ago, portions of the North American highlands may have looked like this picture of depleting ice and emergent land, as the Wisconsin ice sheets retreated. There was no way of knowing the present trend of the Antarctic ice cap, even though the past was plain. Ice retreat in the massif, as in the Sentinels a thousand miles away, could have occurred thousands of years ago. It could have halted or have become reversed.

From the Pensacolas, the expedition turned west. It was Thiel's intention to survey the southern boundary of the shelf, but time was running out. During the period that radio communications were blacked out, Neuburg could see in his theodolite the sunspots causing the disturbance in the radio-wave-reflecting ionosphere. When the blackout lifted, a message came through from Ronne saying: "You have no alternative but to return." This message was overheard by Bentley, who picked it up in Marie Byrd Land.

Thiel and Neuburg then turned north toward the coast. On December 31 Augenbaugh spied an indistinct penguin track at a point the investigators thought was at least two hundred miles from the nearest open water. It was another curious, inexplicable find, made two days before the Sentinel Traverse saw a similar track in the interior of the West Antarctic highland.

Photographs of the track were later shown to Sladen, who believed it had been made by an emperor penguin, the largest of the penguin clan. What the emperor was doing so far in the interior of the Filchner Ice Shelf was as much a mystery as the mission of the little Adélie. The two tracks amounted to a curious coincidence.

On the northward trip the traverse encountered heavy fields of crevassed ice which forced it to halt 200 miles southwest of Ellsworth Station. Ronne was insisting on an immediate return. The *Staten Island* and *Wyandot* had returned to the ice shelf with another season's supplies, and Captain Edwin A. McDonald was holding the ships against the shelf, waiting for the scientists who were scheduled to sail with him to New Zealand and home. By mid-January, after

Scientists strain the waters of Lake Bonney, one of the rare lakes found in Antarctica, which is located in the ice-free region of Victoria Land.

The USS Staten Island *during its operations near the ice shelf.*

becoming more concerned daily about the ice piling up in the Weddell Sea, he couldn't wait any longer. He flew out in an airplane to pick up the research men, and a second aircraft ferried out replacements.

The replacement scientists were the Reverend Father Edwin A. Bradley, S. J., a seismologist of Xavier University, Cincinnati, and John Pirritt, a Scottish glaciologist, who were scheduled to cross the shelf and the West Antarctic highland to Byrd Station the following season.

Thiel, Neuburg, and their colleagues flew back to the station and went aboard the *Wyandot,* while Father Bradley and Pirritt stayed with the stranded Sno-Cats. They strove vainly to run them back to the station, but could not clear the crevasses. Autumn temperatures were falling rapidly. After parking the vehicles for the winter on the shelf about 125 miles from the station, where there was no parking problem at all, the replacements were airlifted back to Ellsworth, where they settled in for the winter.

The first Ellsworth Traverse had covered 1250 miles in 81 days, making three discoveries. It had picked up the trough, had found the shelf divided by an "island," and had demonstrated that the floating ice was three times as large as previously suspected, with an area of 140,000 square miles.

When the night of winter fell in March 1958, the nature of the land beneath the ice was still a mystery, more tantalizing than ever. A trough dividing West and East Antarcticas had been found running inland from the Ross Shelf for 1200 miles. A second trough beneath the Filchner Ice Shelf seemed to go inland to meet the first. It was logical to conclude that there was a connection, that the fabled Ross-Weddell channel did, in fact, exist.

But how could this be? Astride the trough rose the Sentinel Mountains, which seemed to cut it off.

Perhaps the missing linkage lay between the Sentinels and the polar plateau. One way to check out this possibility was to cross the region from the Filchner Ice Shelf to Byrd Station. That was the mission Pirritt and Father Bradley proposed to undertake when spring came in the last quarter of 1958.

The trough vanishes

In the 1958–1959 austral summer, the question of whether the trough in Marie Byrd Land linked up with the trough under the

Filchner Shelf was attacked by three expeditions: Pirritt and Father Bradley led a traverse across the ice from Ellsworth to Byrd Station, passing to the south of the Sentinels: Thiel, Neuburg, and Ostenso made a sequence of seismic soundings across Marie Byrd Land, from the north coast to the south polar plateau, using aircraft to hop from one side to another; and the indefatigable Bentley led an oversnow traverse to the Horlick Mountains, which seemed to form the central segment of the Pre-Cambrian Trans-Antarctic mountain system.

The Pirritt-Bradley Traverse left Ellsworth Station October 31, 1958, on the 1250-mile trek to Byrd Station. A. Goodwin assisted Pirritt in glaciology; F. T. Turcotte, then a graduate student at the University of California at Berkeley, worked with the Jesuit priest in seismology.

The party crossed the Filchner Ice Shelf in the two Sno-Cats which, after having been parked all winter on the shelf, had been towed back to Ellsworth in the spring and reconditioned. East of Berkner Island, the team sounded the trough and followed it south-westward for 250 miles. Where the floating ice shelf merged with the grounded ice of the West Antarctic highland, both the snow surface and the rock bottom began to rise in elevation. From a depth of 4000 feet below sea level, the bottom contour climbed to 4000 feet above sea level, a rise of more than 1.5 miles. In the vicinity of 88 degrees west longitude, the party was riding on an ice surface 5000 feet above sea level. Only 1000 feet below, the seismic readings showed, was rock—a buried mountain range.

At 88 degrees west the trough had vanished. On the featureless snow surface the geographical location was devoid of meaning. But if one followed this meridian north of the party's position, one came to the Sentinel Mountains. Southward lay the polar plateau and the Horlick Mountains, part of the Trans-Antarctic Range. Thus the expedition was passing between the Andean Sentinels and the Pre-Cambrian mountain rim of East Antarctica. It appeared that the high rock represented a southward extension of the Sentinels under the ice. Pirritt and Father Bradley determined that the buried mountains ran at least 70 miles across the projected axis of the trough they had traced on the ice shelf.

Rolling on to the southwest, the party cleared the sub-ice mountain region. Once more the rock bottom sank below sea level and the traversers found themselves over the sea-bottom channel of Marie Byrd Land, which they followed to Byrd Station.

This traverse ended fifty years of speculation that a broad channel connected the two embayments of the Ross and Weddell Seas. If any channel did cut through the 88th meridian south of the expedition's track, it was nowhere near the magnitude of the trough in Marie Byrd Land.

Now there remained to be determined how wide the trough was in Marie Byrd Land from north to south, or from the volcanic mountains of the coast to the Archean battlements of the polar plateau. This job was done by Thiel, Neuburg, and Ostenso, who began an airborne survey of the width of the trough on December 8, 1958, and completed it by New Year's Day, 1959.

From Byrd Station, at the junction of the 120th meridian and the 80th parallel, they planned to sound the ice at seven sites along the meridian between the north coast and the polar plateau. It was feasible to use an airplane for this kind of spot checking in Marie Byrd Land, for West Antarctica was lower in elevation and warmer than the 8000-foot plateaus of East Antarctica, where airlifted reconnaissance had proved dangerous the year before. The three investigators were given the use of the Navy R4D airplane named the *Takahe*, which had ferried supplies to the Bentley-Anderson expedition to the Sentinels in the previous season.

In all, the survey required seven landings as planned along a north-south line of 400 miles. The pilot, Ronald Carlson, and his crew made runways at the sites by taxiing the aircraft back and forth over the soft snow on its skis. At all seven stations the ice was found to be resting on rock, which dismissed any lingering notion that there might be water in the channel. The trough ranged in depth from 1640 feet below sea level near the coastal mountains to 3280 feet below sea level near the ranges of the plateau.

The slope of the ice surface was measured. Instead of sloping down to the sea from the plateau, it sloped downward to the plateau from the sea—the reverse of what had been expected—demonstrating that the ice of West Antarctica was flowing not into the Pacific Ocean, but into the Ross Ice Shelf.

Between the coastal mountains and plateau, the ice formed a 400-mile concave saddle, which reflected the broad contour of the trough beneath. On this scale the ice did follow the contour of the land beneath. Early explorers had supposed that it did, but on a much smaller, more localized scale. When the ice surface rose in elevation, they had imagined that the land beneath it rose too. Thus Amundsen and later Byrd had surmised rising land from the edge of the Ross

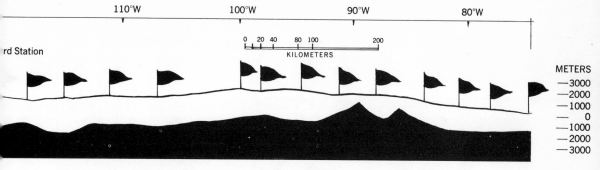

Land profile beneath the ice, according to findings of the Ellsworth-Byrd traverse, 1958–1959.

Shelf, because the ice sloped upward as though defining some continental shore. It did not seem reasonable then to suspect that the rising ice of Marie Byrd Land concealed a deepening trough.

On the enlarged scale, running from the coast to the polar plateau, the apparent concavity of the 400-mile-wide Marie Byrd Land ice sheet might have suggested the trough that lay beneath, but this feature was so massive it was not visible to men on the ice, or even from the air.

The third assault on the problem of the trough in the 1958–1959 season was Bentley's 300-mile probe due south of Byrd Station to the Horlick Mountains. The range, which had been sighted on an airplane flight of the Byrd expedition in 1935, was named for William Horlick, the malted-milk-maker, who was one of the expedition's financiers.

As far as the sub-ice construction of West Antarctica is concerned, the significance of the Horlick Mountains Traverse is that it covered the region between the route of Pirritt and Father Bradley and that of the airborne survey. With Bentley were the Long brothers, William and Jack; Leonard LeSchack, a seismologist; Frederick L. Darling, United States Weather Bureau meteorologist; and William H. Chapman, a topographical engineer of the United States Geological Survey. In addition to sounding the ice across Marie Byrd Land, the traverse proposed to examine the mountains and ascertain whether they were part of the same chain as the ranges rearing up along the Ross Sea coast of Victoria Land in East Antarctica.

The expedition left Byrd Station on November 1 in three Sno-Cats and sounded the trough along the southward track for 340 miles. The land rose to sea level only when the party approached the mountains.

Like the Sentinels, the Horlicks were guarded from the encroachment of man by a vast moat of crevassed ice. Turning east, the expedition cruised on a course parallel to the range for 200 miles.

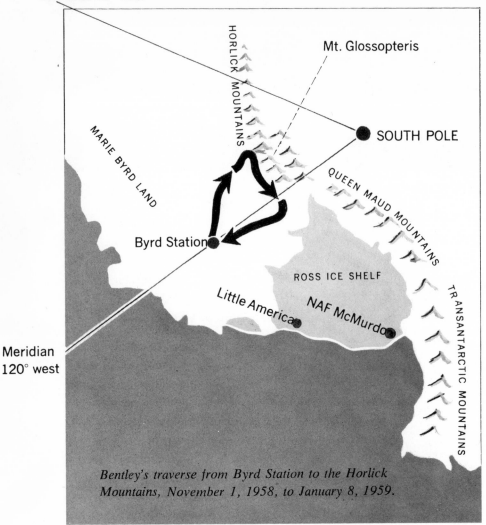

Meridian 90° west

Mt. Glossopteris

SOUTH POLE

HORLICK MOUNTAINS

MARIE BYRD LAND

QUEEN MAUD MOUNTAINS

TRANSANTARCTIC MOUNTAINS

Byrd Station

ROSS ICE SHELF

Little America

NAF McMurdo

Meridian 120° west

Bentley's traverse from Byrd Station to the Horlick Mountains, November 1, 1958, to January 8, 1959.

George A. Doumani, a Lebanese geophysicist, and Fien Chang, a Chinese seismologist, were flown out to join the party.

It was during the eastward leg of the traverse that Bentley, the Long brothers, and Chapman climbed a mountain peak and found new evidence of the lost world of a warm Antarctica in the prehistoric past. The discovery was to rekindle speculation about the nature of the land in previous epochs, as we shall see.

Turning northwest, back toward Byrd Station, the party sounded a deep canyon that plunged 3200 feet below sea level, though it was apparently narrow. Thirty miles beyond it, the sub-ice surface rose to 6560 feet above sea level. Was this gorge the linkage between the West Antarctic and Filchner channels? It was tempting to think so, but the investigators had become wary of drawing conclusions from

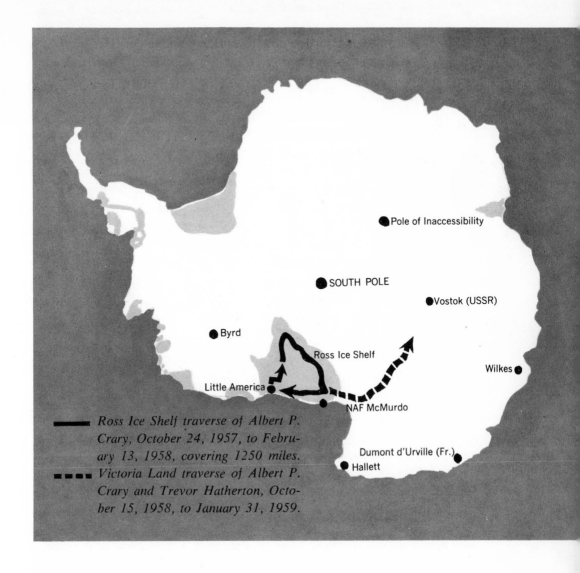

Ross Ice Shelf traverse of Albert P.
Crary, October 24, 1957, to February 13, 1958, covering 1250 miles.
Victoria Land traverse of Albert P.
Crary and Trevor Hatherton, October 15, 1958, to January 31, 1959.

limited data. Bentley believed such a surmise on the basis of one sounding would not hold up very well. It was more likely, he thought, that the deep represented a large fjord.

Moving westward, the vehicles once more rolled over the great trough. On December 29, Lieutenant Commander Robert Epperly reached the expedition in a resupply airplane and advised the scientists that Little America, the historic base in the Antarctic, was going to be closed. Scientists who planned to return to the United States were to assemble at McMurdo Sound, where the *Wyandot* would pick them up and take them home.

Most scientists who know the region regarded Little America V as the best scientific base this country had on the ice. But the Navy ruled that the base should be closed and its functions as a port of entry and science station should be consolidated with those at Mc-Murdo Sound, on the west, and Byrd Station, to the east. Aside from the question of cost, Navy spokesmen said at the time that the ice on which Little America had been constructed was becoming unstable. The station would have to be moved anyway to another part of the Ross Shelf. The closing of Little America also closed a famous landmark and ended a long Antarctic tradition. To more than a hundred scientists and Seabees, Little America had become a home away from home. Many of them protested its closing in letters to Congressmen. But the Navy stuck to its guns, and Little America V became, like its predecessors, a memory.

The order meant that the party had to hurry back to Byrd Station to catch a special airplane flight across the Ross Shelf to McMurdo Sound on January 8, 1959. And so the last major traverse of the International Geophysical Year in West Antarctica ended, within a few days of the formal close of the IGY on December 31.

Only one United States expedition remained out on the ice. Crary was still journeying on the high plateau of Victoria Land in East Antarctica. With him were Dr. Trevor Hatherton, chief scientist of the New Zealand Antarctic program, Charles R. Wilson of the United States Geological Survey, Lyle McGinnis of the United States Weather Bureau, Stephen Den Hartog, a glaciologist, and Frank Layman, a mechanic.

Beneath the high ice of the East Antarctic plateau the scientists had sounded a rocky plain lying at sea level. Except for the ancient mountains bordering East Antarctica, the predominant continental configuration appeared to be an enormous plain. Some of it sounded

Stephen Den Hartog of the University of Wisconsin measuring the earth's gravity on an island near McMurdo Station.

below sea level, but Crary estimated that, if the weight of the ice was removed, the land would rise to sea level or above, as a result of being freed of its icy burden.

Sixty years after John Murray had presented his evidence of a continental Antarctica to the Royal Society, the hypothesis had been proved. Soundings by Crary in Victoria Land, by Fuchs on his crossing of the East Antarctic ice cap from the Weddell to the Ross Seas, and by Russian, French, and Australian seismologists all demonstrated that East Antarctica was a continent larger than Australia.

What of West Antarctica on the cone of the bell? Much of it had turned out to be sea bottom. This seemed to explain why the measurement of earthquake waves on trans-Antarctic paths indicated that only three-fourths of the Antarctic ice sheet was underlain by continent, with the remainder oceanic in structure. The velocities of these waves suggested that the average thickness of the crust in Antarctica was three-fourths that of a continent.[5]

Sea bottom versus continental land

When they returned to the United States after the IGY, Bentley, Crary, Ostenso, and Thiel drafted a joint report on their findings in West Antarctica.[6] Instead of linking the Ross and Weddell Seas, as had been speculated, the channel seemed to veer northwest of the Sentinel Mountains toward the Bellingshausen Sea. Beneath the West Antarctic cap, therefore, the scientist could visualize an oceanic region consisting of mountainous islands on the north, a frozen sea comprising most of Marie Byrd Land, and the extension of the Antarctic Peninsula into the heart of the West Antarctic highland.

If the ice were removed, the continent would lie some 400 miles south of what now appears to be the north coast of Marie Byrd Land. The coastline would consist merely of large islands, almost as far from the mainland as the Bermudas are from the Carolinas. A ship might sail into what we think of as Marie Byrd Land all the way south to the Horlick Mountains, 300 miles from the south pole. What now appears to be the Antarctic Peninsula may well be an island, long, narrow, and mountainous. If the Sentinels are a part of it, the great island must continue southward a thousand miles beyond the present coastline. Perhaps it is separated from the continental barrier by a narrow channel, but certainly not by a broad one. Compared to the contour of the ice cap, the true Antarctic continent would appear greatly shrunken if the ice were gone.

As for the ice itself, Bentley, Crary, Ostenso, and Thiel thought the West Antarctic sheet originated as two separate ice caps, formed in the mountains north and south of the pre-glacial sea. According to this theory, in the north the ice accumulated atop the volcanic coastal ranges, and in the south it built up in the Trans-Antarctics. Then the mountain ice caps flowed across the intervening sea and merged into the single sheet we see today. Ice is a plastic material which moves imperceptibly under the pull of gravity. Since the flow today is demonstrably toward the Ross Ice Shelf, it is reasonable to speculate that the shelf was formed in early times by an overflow from West Antarctica. And, as the ice front of the shelf rises today in about the same position as in the time of Scott and Amundsen, we can assume that whatever changes are taking place in the West Antarctic cap are not very rapid—certainly not so rapid as climatic changes occurring in our own part of the world, the northern hemisphere, might suggest—although this is another story we shall discuss later.

At the end of the IGY in 1959, the picture of the land beneath the ice was considerably more detailed than in 1910. Not only had Murray's hypothetical continent become a reality, but the great trough and the divisions of Antarctica had been identified. However, the picture was still incomplete. Where did the channel go from the Sentinel Mountains? Was the Panhandle a peninsula or an island? Did it belong to South America or to Antarctica? Where was the junction or separation of the Andean and Antarctican mountain systems? What did such a configuration represent in the evolution of the earth?

Far from resolving the Antarctic mystery, the IGY effort had merely got the bear by the tail. The scientist could not let go. More than ever was he impelled to continue the investigation of Antarctica. Otherwise, the enormous investment of men and money during the IGY would have lost much of its value.

By general agreement, the IGY was followed in Antarctica by a regime of International Geophysical Cooperation. It represented an indefinite continuation of the international investigation, subject to the political moods and economic abilities of the participating nations.

Post-IGY begins

The IGY had shown the government of the United States that years of research lay ahead. The American policy, as defined by President Eisenhower in 1958, was to get on with it. A similar policy was being followed in the Soviet Union. There is little doubt that each influenced the other.

Administratively, there were changes. The National Science Foundation assumed the responsibility for the Antarctic program on a long-term basis. It created an Antarctic Research Program in its Office of Special International Programs to manage the investigation, and Dr. Thomas O. Jones was appointed director of the program in 1959, and Crary was named chief scientist. Then, in 1961, the Antarctic Program was upgraded as an "Office" of the foundation.

There was also a change in the investigative method after the IGY, during which the exploratory and other scientific programs had been mapped out by a special committee which then recruited scientists to carry them out. When the foundation took over Antarctic research, it allowed freer reign to the universities, whose scientists could follow their own research interests with federal support, within limits prescribed by the foundation and advisory groups.

Thus, under the post-IGY program, no detailed research plan was formulated. University and other research organizations were encouraged to devise their own projects, within the foundation's general scheme of broad research objectives. In geophysics, as in the other disciplines, the major questions were obvious. No one in Washington had to tell the University of Wisconsin, or Ohio State University, or any of the other institutions engaged in the investigation what to do.

Left to right: *Dr. Thomas O. Jones, Sir Vivian Fuchs, and Admiral David A. Tyree, about to enter the shaft leading to the new Byrd Station, which was built under the snow surface and completed in 1962.*

The institutions submitted proposals and those accepted by the foundation were financed by federal grants.

When American scientists returned to the ice in the last quarter of 1959, they found that the United States had tightened its belt for the long haul ahead. Of the original Seven Cities, only NAF McMurdo, Byrd, and Amundsen-Scott Stations remained under full United States control. Hallett Station at the entrance to the Ross Sea was turned over to New Zealand for administration and supply, Argentina assumed the logistical responsibility for Ellsworth, and Australia did the same for Wilkes on the Indian Ocean side of East Antarctica. American scientists continued to man the stations together with scientists of the supporting countries.

Meanwhile, the guard was changed at McMurdo Sound. Dufek was retiring from the Navy. He was replaced as commander of Operation Deep Freeze and Task Force 43 by Rear Admiral David M. Tyree. Dufek had presided over the construction of the bases and had written his name in Antarctic history. He was a popular figure in New Zealand, where relations between Navy personnel and the civilians have always been extraordinarily good. Admiral Tyree continued that tradition.

The story is told that shortly after Admiral Tyree arrived at McMurdo Sound he wandered into the enlisted men's "head" for a shave. It was only after he had donned his parka and cap to leave that he was recognized by a young power-shovel operator who had been washing next to him. The youngster gaped with surprise.

"What's the matter, son?" asked Tyree. "Haven't you seen an admiral shave before?"

"No, sir," said the sailor. "I never even saw an admiral before."

The physiographic question in 1959 was whether a break existed between the Andean Sentinels and the Trans-Antarctic Mountains rimming the polar plateau. A broad trough in that region had been ruled out, but perhaps there was a narrow gulf, as suggested by Bentley's deep sounding on the Horlick Mountains Traverse.

The job of hunting for such a break fell to Thiel, who returned to Byrd Station in October 1959 to search for it in a series of airlifted soundings along the 88th meridian, together with the geologist Campbell Craddock, and the seismologist Edwin Robinson, of the University of Minnesota.

At their disposal was the Navy's venerable airplane *Takahe,* piloted by Carlson, who in the previous summer had flown Thiel and Ostenso

Dog team for a field party of four New Zealand geologists, with a Navy R4-D airplane in the background, at the camp site of the scientists.

to sites along the 120th meridian. Like a number of Navy pilots in the Antarctic, Carlson had become as fascinated with the physiographic problems as the scientists. Consequently the Navy VX-6 (Very Exceptional Six) Squadron was flying this mission far beyond the normal search and rescue range of aircraft based at McMurdo Sound.

The plan was to make eight landings between the Sentinels and the Eastern Horlicks. Because of the deep canyon that Bentley had sounded near the Horlicks, Thiel expected that a gap might be found fairly close to the plateau mountains.

As the airborne traverse worked its way south, flying from one site to another, it reached the southern boundary of Antarctandean rocks. South of this point, the nunataks rising up through the snow were not the folded metamorphic rocks of the Sentinels but the block uplifts of Pre-Cambrian Horlicks.

It was beyond the Sentinel system that the party sounded rock surface below sea level. There was a gap, but at first the calculations showed it did not run deep. If the ice cover were removed, the rock surface, relieved of its weight, would rebound to an elevation of about sea level. But farther south, nearer the Horlicks, the soundings began to show greater depths. The rock bottom dropped to 4920 feet below sea level. Allowing for the rebound of the crust with the ice removed, the channel would still be a half-mile deep.

Thiel and his colleagues concluded that a trough 30 miles wide divided the two mountain systems. It was evidently a comparatively narrow gulf or strait, on the order of the English Channel between Dover and Calais. It could be linked by extrapolation with the deep channels that Thiel had found beneath the Filchner Ice Shelf and that Crary had sounded beneath the Ross Ice Shelf. Did this narrow channel run across the bell? Was it an offshoot of the main trough in Marie Byrd Land? Or was it a separate split in the crust?

Bentley returned to the Antarctic in 1960 to explore the region between the base of the Panhandle and the Bellingshausen Sea. The expeditions of 1958–1959 had shown that the Byrd Land trough was blocked in the direction of the Weddell Sea by the Sentinel Mountains. Now Bentley proposed to determine whether it went to the west of the Sentinels into the Bellingshausen Sea, as he and others had suspected.

On November 14 he set out from Byrd Station with a team of scientists in three Sno-Cats. The expedition made a zigzag trail through the Ellsworth Highland between the Sentinels and the coast. North and west of the Sentinels, the soundings showed great deeps. The ice was 8200 feet thick and the land beneath lay more than a mile below sea level. Unquestionably, the trough trended to the Bellingshausen. But as the party approached that forbidding coast, on which no man had ever before set foot, the investigators were astonished to find that the land rose to sea level!

Now what? The trough had vanished again. It did not go to the Bellingshausen Sea at all. Instead, that coastal region under the ice seemed to represent a westerly extension of the Panhandle.

Working eastward from the coast, the expedition sounded a channel which seemed to trend toward the base of the Panhandle in the direction of the Weddell Sea. Then, turning back and probing westward, the scientists found a channel in the direction of the Amundsen Sea. Now it appeared that the West Antarctic trough made a Y branch somewhere between the Sentinel Mountains and the sea. One branch ran toward the Amundsen Sea and the other toward the Weddell through the base of the Panhandle.

These indications upset the concept of the structure of West Antarctica that had evolved since 1957. If a channel cut across the base of the Panhandle, it meant that the Antarctandes were separated from their supposed extension, the Sentinels. If a second branch channel ran to the Amundsen Sea, it suggested that the Bellingshausen sea-coast was actually a part of the Panhandle.

West of the Sentinels, the picture of West Antarctica that scientists had visualized fell apart. No one now could figure out this maze without further investigation. It was becoming more and more probable that there was no such thing as West Antarctica under the ice cap, but simply an assortment of islands, an Antarctic Oceania.

Instead of clarifying the problem of the great trough, the 1960–1961 expedition increased it by revealing new and unsuspected complexities. In this respect the expedition confirmed the "law of the second look," which states that the more closely one examines an Antarctic region the more complicated it becomes.

Much of the Bellingshausen Sea region from the Eights Coast to the Robert English Coast had been mapped by aerial reconnaissance of Finn Ronne in 1947. But when Bentley and his colleagues took

A Neptune P2-V flying over Beardmore Glacier en route to the south pole.

a second look in 1960–1961, some of the landmarks were missing. Two mountains, Mount Tuve and Mount Peterson, were "gone." They had been marked on the maps of the National Geographic Society for years. Yet the expedition did not find them when it reached the sites where the map said these lofty landmarks, 9000 feet high, were supposed to be.

Bentley and his associates camped on the site of Mount Peterson, a flat snowy plain. They were unable to see Mount Tuve as their Sno-Cats rolled over the coordinates marking its reported location.

These experiences are not unusual. Navy pilots have reported flying through mountains marked on their charts. Mapping the surface is an inexact business in the Antarctic, especially when it is done from an airplane.

Flying to the south pole in November 1960, I remarked I could see bare rock below as the aircraft, a Lockheed Hercules C-130, passed over the top of the Beardmore Glacier.

"There damn well better be," muttered the navigator, "or I wouldn't know where we are." He was checking position from the naked peaks of the Trans-Antarctic Mountains.

It is not the mountain which gets lost in Antarctica, but the man. Yet the difficulties of mapping the ice surface are mild compared with those of charting the land beneath it.

After three months of probing the coastal region, Bentley's party returned to Byrd Station on February 12, 1961. What had been proved was that further investigation was essential before the problem of the land beneath the ice of West Antarctica could be solved.

It seems strange that this region of the earth continues to defy investigation in a day when we are talking about the manned exploration of the moon and have already sounded shrouded Venus. But the investigation of Antarctica has only started. Most of the American work has been in West Antarctica. The job of making detailed soundings on the continental cap of East Antarctica is still ahead.

Beyond this lies an even more fascinating mystery. We know that, long ago, all these lands were fair and open to the sky because there is coal in the Trans-Antarctic Mountains. Where there is coal there was once a forest. And where a forest grew, we must assume that the climate was temperate, semi-tropical, or even tropical.

GONDWANALAND

Expeditions to the central Horlick Mountains of the polar plateau in 1958 and 1960 found new and striking evidence that ice-capped Antarctica once basked in a temperate or tropical climate. The evidence included large fossil tree trunks, thick seams of anthracite coal, and the fossil fern *Glossopteris*. Fossils of this prehistoric fern have been found in South Africa and India.

Of primary importance, the tree stems and the coal prove that a rather respectable forest once flourished in this region, which nowadays is less than three hundred miles from the south pole and is immersed in the long polar night nearly six months out of every year. The implication of this evidence is startling. It tells us that Antarctica must have had a more favorable relationship to the sun in bygone eons than it has today. It is hard to see how a lush forest could have grown in a region with six months of darkness, irrespective of the climate.

If Antarctica's position with respect to the sun has changed, that means that either the continent has moved or the earth's axis of rotation has wandered. For, in a land which has a coal forest at one period and an ice age at another, something has had to give. For reasons we shall examine shortly, the findings in the Horlick Mountains lent weight to the idea that the continent has migrated, thus supporting the strange, controversial theory of continental drift.

125

The hypothesis of continental drift and sliding has been in and out of style, like double-breasted suits, for half a century. It was first proposed in 1912 by Alfred Lothar Wegener, professor of meteorology and geophysics at the University of Graz, Austria. Wegener was one of those rare observers who did not hesitate to base a radically new idea on observations that were obvious to everyone. His theory of the origin of the continents and the oceans is founded in part on this observation:

He who examines the opposite coasts of the South Atlantic Ocean must be struck by a similarity of the shapes of the coastlines of Brazil and Africa. Not only does the great, right-angled bend formed by the Brazilian coast at Cape San Roque find its exact counterpart in the re-entrant angle of the African coastline near the Cameroons, but also south of the corresponding points, every projection on the Brazilian side corresponds to a similarly shaped bay on the African side and vice versa. Experiment with a compass and globe shows their dimensions agree accurately.[1]

Wegener was not the first scientist to comment on apparent matching of these coastlines. Similarities of the western coastline of Europe and Africa to the eastern coastline of North and South America were noted in 1620 by the British philosopher and scientist Francis Bacon. He suggested they seemed to lock together like the pieces of a gigantic puzzle. The similarities also were noted by the French zoologist Georges Buffon in 1780, and by the American astronomer Edward C. Pickering in 1880.

Three phases of the disintegration of Pangaea into continental fragments, according to Wegener's hypothesis.

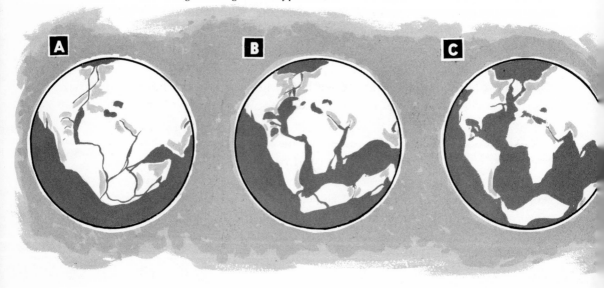

But Wegener was the first to use the observation as the basis of a theory about the evolution of the earth. He hypothesized that in the beginning all the continents were joined together into a supercontinent called "Pangaea." Slowly, in response to forces within the earth, portions of Pangaea drifted off and have continued to drift over millions of years. If one thinks of the continents as pieces of a terrestrial jigsaw puzzle, one begins to examine the coastlines and finds that these pieces do appear to fit together. Of course, the fit is approximate, but the continental shelves must be taken into consideration.

In the southern hemisphere, according to Wegener's theory, Australia, India, Africa, and South America were all wrapped up together, with Antarctica in the center. The northern end of this huge ball consisted of Europe joined to North America through Greenland, Iceland, and Ireland. Why do the Appalachian Mountains stop dead at the coast of eastern Quebec and Newfoundland? Are they the same as the Caledonian Mountains of England and Scotland, and the mountains of Norway? Wegener contended this mountain system was pulled apart by the disassociation of North America and Europe. Geological evidence indicates the American and European mountains are the same age and have undergone the same folding. And, of course, they look the same—a fact which may help account for the settlement of Appalachia in the eighteenth and nineteenth centuries by English and Scottish immigrants.

There is all manner of other evidence for Wegener's theory—and against it, too. The idea of a supercontinent was not original with Wegener. In a more modest way it had been proposed in 1885 by Eduard Suess, a London-born Austrian geologist noted for a classic work on the Alps. Suess was impressed with similarities observed by himself and others in both the fossil and modern flora of South Africa, Madagascar, and India. He supposed these lands had been connected in a supercontinent he called "Gondwanaland."[2] The term, with a fine Afro-Indian ring, is derived from an ethnic group in India called the "Gonds." Suess wrote in 1905:

The southern and a great deal of the more central parts of Africa, Madagascar and the Indian peninsula comprised one great continent. We call this mass Gondwanaland after the ancient Gondwana flora which is common to all its parts. This country is incomparably older than North America. Indo-Africa is the great table land of the earth.[3]

But since Suess began to connect up the land masses, a great many new observations have been made. The characteristic fossil flora of Suess's Gondwanaland was the fern *Glossopteris*. The finding of this fossil plant in the mountains of Antarctica places the seventh continent in the Gondwanaland complex.

Though there never has been any general acceptance or rejection of the Gondwanaland hypothesis, it has grown on scientists over the years and many of them use the term as though they fully accepted the theory. When paleobotanists or paleontologists speak of "Gondwanaland flora" they mean plants found in all or most of the southern-hemisphere lands. In modern parlance, Gondwanaland refers to all the lands of the southern hemisphere, including Antarctica. And whether one puts any stock in the Gondwanaland theory or not, the usage of the term is steadily increasing as more and more geological, botanical, and faunal links between the widely separated land masses are found.

But when the concept of Gondwanaland is stretched to Pangaea, then the argument becomes heated. For a number of prominent geophysicists, geologists, and oceanographers are convinced that the continental blocks cannot move horizontally through the rigid basalt structure of the ocean floor, as the continental-drift theory implies.

But whether continents can drift or not, the great Gondwanaland linkage, *Glossopteris,* is an indisputable and incontrovertible fact.

The linkage of life

The botanical connection between the continents of the south has been observed for more than a century. In 1847 Dr. Joseph Hooker, the surgeon with Sir James Clark Ross's seafaring expedition to McMurdo Sound, wrote:

> The many bonds of affinity between the three southern floras—the Australian, the Antarctic, and the South African—indicate that they may all have been members of one great vegetation, which may once have covered as large a region in the south as the European flora covers in the north.[4]

Similarities of the modern vegetation in the widely separated lands of the southern hemisphere have aroused the curiosity of many scientists. Dr. Martin Holdgate of the Scott Polar Research Institute, Cambridge, England, theorizes that plants and some animals may have been dispersed through Antarctica, when its lands were ice free:

> When a New Zealand botanist enters the evergreen rain forest that blankets the western hills of Chiloé Island in southern Chile, his first impression is of being in a familiar environment. Broad-leaved mixed woodland alternates with forest of the southern beech much as it does in New

Zealand 5000 miles away across the South Pacific, and the dominant trees belong to genera which are also important in temperate Australasia: *Eucryphia, Weinmannia, Laurelia, Nothofagus,* and *Podocarpus*.* Out of 46 families present in these Chilean forests only seven are unrepresented in the New Zealand equivalents, while 40 per cent of the genera are shared and have a similar habit of life in the two regions. . . . In the moss of the Chilean beech forest tiny flattened Hemiptera [an order of insects usually called bugs] belonging to the primitive family *Peloridiidae* are locally common, just as they are in New Zealand, Tasmania, the temperate forests of the Australian mountains and Lord Howe Island—but nowhere else in the world.[5]

The earliest evidences of Antarctica's warm past were the fossil ferns, mollusks, and coniferous wood found on Antarctic islands in 1893 by Captain C. A. Larsen, master of the sailing ship *Jason*. More fossils were found at Admiralty Inlet on the coast of the Panhandle by F. W. Stokes, an artist on the *Belgica* expedition to Antarctica between 1899 and 1902. Stokes meticulously sorted and classified the fossils and brought them to the United States. His collection was analyzed in 1903 by Professor Stuart Weller, a paleontologist of the University of Chicago, who identified three species of snail-like gastropods identical with those found in southern India. Weller also identified among the fossils two other gastropod species, which had also been found in the Strait of Magellan.

So it was that the linkage between Antarctica and other southern-hemisphere continents had been in evidence about the turn of the century through fossil plants and animals. More evidence—much more—was yet to come.

Hartley T. Ferrar, a geologist with Scott's first British National Expedition of 1901–1904, found *Glossopteris* fossils and coal seams in the sandstones of the Prince Albert Mountains of Victoria Land. This range is from 42 to 60 miles across McMurdo Sound from Ross Island, though on a clear, sunny day it seems only a dozen miles away. The great, rearing battlements of naked rock and gleaming ice are one of the grand sights of the earth.

The first tree-stem fossils were found during the Scott expedition by Sir Raymond Priestley, who reported seeing tree stems 12 to 18 inches in diameter in the sedimentary rocks of the great barrier of

---

* *Eucryphia* is a genus of tall evergreen tree found in Chile and Australia; *Laurelia,* known as Peruvian nutmeg, is a timber tree also found in New Zealand; *Weinmannia* refers to a genus of shrubs and trees with simple leaves and clusters of flowers, found throughout the southern hemisphere; *Nothofagus* is a genus of timber trees growing in the far south; and *Podocarpus* is a genus of evergreen trees belonging to the yew family and sometimes called black pine in Australasia.

Victoria Land. Later in the decade, Shackleton and members of his party found seams of coal embedded in the sandstone walls which form the "banks" of the gigantic Beardmore Glacier. And, of course, there was Dr. Wilson's impressive collection of specimens, left outside the tent on the Ross Shelf where Scott and his men had died in 1912.

Members of the 1935 Byrd expedition found *Glossopteris,* fossil tree trunks up to 18 inches in diameter, and fifteen seams of coal, interbedded with shales and limestones, in the Queen Maud range of the Trans-Antarctic Mountains, the same mountain system which runs along the coast of Victoria Land. There seems to be little question now that it extends all the way across Antarctica from the Pacific Ocean to the Filchner Ice Shelf on the Atlantic Ocean. Nearly all the major fossils finds reported have been discovered in this mountain system.

When Bentley and his associates approached the Horlick Mountains in their Sno-Cats on the long trek out of Byrd Station in 1958, the party made camp on December 7, three miles from a peak which reared up majestically from the surrounding snowscape. Two days later Bentley, Bill and Jack Lang, and Frederick L. Darling began to climb it, and Bentley reported the event in his diary:

There were low clouds over the peak. The first half of the climb was through them. This part was all rock, with many and varying, interesting geologic features, including coal beds, petrified wood, and fossils near the top and at the bottom of the sedimentary section.

We reached the top of the clouds and of the bare rock at about the same level. After waiting a few minutes, the crest of the peak gradually appeared through the lightening mist, a very beautiful sight.

The peak reached an elevation of 4000 feet above sea level. In one 1860-foot-thick section of the sedimentary layers they saw sandstone, shale, and coal beds repeated several times. Embedded in the sandstones were fossils of the fern *Glossopteris,* so they named the peak Mount Glossopteris.

The climbers found fossils of mollusk-like marine animals and counted 30 beds of anthracite coal, each 3 to 4 feet thick. High above the encircling ice they found a tree, 12 feet long and one foot in diameter, locked in the sandstone. By digging, chopping, and sawing, they succeeded in extricating a foot-long chunk of mineralized wood. Months later, at Ohio State University, it was identified as *Dadoxylon,* a tree well known from its fossilized remains in other lands of the southern hemisphere.

A year later Dr. Laurence M. Gould, dean of American Antarctic

Ohio State University geologists gaze up at the north ridge of Mount Weaver in the Queen Maud Mountains, 220 miles north of the south pole.

scientists and chairman of the Antarctic program of the United States National Committee for the International Geophysical Year, told an appropriations subcommittee of the United States House of Representatives:

As a geologist, I must single out one exciting discovery of the Byrd Station Traverse during the past summer season. Dr. Charles Bentley and his team journeyed south until they reached the northern edges of the Horlick Mountains. There, to their amazement and delight, they found a tremendous store of fossils: a petrified tree trunk 12 feet long, leaf fossils, fossil bivalve shells and coal beds.

All through the years that fossils and coal have been turned up in Antarctica, their implication has stirred the imaginations of the best scientific minds in the world. Fossils of tropical plants also have been found in the Arctic, in Greenland, and Spitsbergen. The coal mines on Spitsbergen testify that in Permian times, some 250,000,000 years ago, forests were growing as near to what is now the north pole as the Antarctic coal beds are to the present-day south pole.

There is no doubt that in past times the two regions were warm and open to the sun. But what was the reason for this climatic change? One widely accepted theory states that the earth was warm for long periods of time from pole to pole, and attributes the higher temperature to more solar energy's reaching earth. The astrophysical explanation excludes such concepts as Pangaea, Gondwanaland, and polar migration. According to this view, the continents have always been more or less where they are today in relation to each other, and the geographical poles have been relatively static. If that has been the case, is it likely that forests could have flourished in polar regions where darkness prevails for six months of the year?

This question was raised in the report, published in 1914, of Scott's final *Terra Nova* expedition. The British scientists who wrote the report pointed out that the concentrated activity of plants growing in Arctic lands compensates for the brief growing season and the "substitution of a continuous period of darkness for daily recurring periods of darkness is not a serious obstacle to the existence of perennial plants." Hence, while a climatic change could be inferred from the evidence of coal and fossil trees, a change in the relationship between Antarctica and the sun could not. But, as an afterthought, it was admitted that sufficient experimental data did not exist to demonstrate the effect on vegetation of a prolonged sunless period irrespective of the climate.[6]

There are no polar forests in our time. The tree line ends where the warmest month is below 50 degrees Fahrenheit, and the only trees man has found in the Antarctic are the mineralized fossils of the Trans-Antarctic Mountains. We live in a rather cool period, assumed by some geologists to be an interglacial one.

In the austral summer of 1960, an Ohio State University expedition returned to a segment of the central Horlicks, now called the Ohio Range. Two members of the party were Bill Long and George Doumani, both research associates at the University's Institute of Polar Studies. Doumani, a Lebanese who had completed his undergraduate work in 1947 at Terra Sancta College in Jerusalem, also had obtained graduate degrees in geology and paleontology from the University of California. Long and Doumani are both "drifters"—supporters of the continental-drift theory.

In the Ohio Range sandstones they found primitive clam and snail fossils, as well as impressions of woody plants and plant spores. These fossils suggested, they pointed out, "an environment of shallow, warm waters, with a sandy, slightly muddy bottom." Such an environment, they added, was characteristic of Devonian times in the region.

The Devonian Period is a 50,000,000-year chapter in the upper

George Doumani, leader of a geological party from Ohio State University, breaks off a rock specimen from a boulder in the Queen Maud Mountains.

Paleozoic Era of the earth's long history. As geologists reckon time, it ended about 305,000,000 years ago and was followed by the Carboniferous Period, lasting 85,000,000 years, when much of the earth's coal was formed. The Carboniferous was, in turn, followed by a 25,000,000-year period called the Permian, with which the Paleozoic Era ended.

The record of the rocks reveals that amphibians, fishes, and invertebrate land animals lived during the Paleozoic, but the giant reptiles had not shown up yet. Their day came in the Mesozoic Era, which followed the Paleozoic. Compared with these vast spans of time, man's own history is barely an instant in the story of life on earth.

Within a 300,000,000-year period many kinds of climate could have come and gone in Antarctica—ice ages interspersed with temperate and tropical times—and on the 1960 traverse to the Horlicks, Long and Doumani found evidence that this was so.

The Devonian rocks were overlaid with rubble called tillite, which is swept up and pressed into a mass by an advancing glacier. The tillite indicated that, after millions of years of warmth in the Devonian Period, Antarctica became cool during the Carboniferous, as an ice sheet, continental in size, crept over it. Far above the tillite Long and Doumani found fossils of higher plants with stems and leaves, showing that Antarctica had warmed up again in the Permian Period, about 220,000,000 years ago.

It was easy to identify the Permian by numerous fossils of the fern *Glossopteris,* characteristic of that period in the southern hemisphere. The tongue-shaped leaf of *Glossopteris* appeared in the same horizontal plane as coal beds, 13 feet thick. The abundance of coal and fossil flora suggests "a lush, green vegetation in a humid, swampy environment," the research men reported.[7]

They found embedded in the sandstones large tree trunks 24 feet long and 2 feet thick, with prominent growth rings. These evoked "a picture of a fast-growing, temperate-zone rain forest not unlike that of Washington and Oregon." The sediments in which the fossils were buried appeared to have been washed into the region from the south —from the direction of the pole, if indeed the pole lay in that direction when the forests grew.

Long and Doumani believed their finds in the Ohio Range were significant in supporting the idea that the continent had moved—at least since Permian times. They pointed out that the juxtaposition of the Permian deposits, with their rich store of fossils, on top of the glacial tillite was not unique. Similar "associations" have been turned up in India, Madagascar, South Africa, South America, New Zealand, and Australia. Such Gondwanaland deposits and other "parallels in the geological record of the land masses of the Southern Hemisphere suggest that much closer ties must have existed in the past." The Gondwana association, they said, is dominated by *Glossopteris* flora. It is likely the southern hemisphere at one time consisted mostly of land, rather than of ocean.

In other southern hemisphere lands Gondwana deposits show a characteristic group of reptiles, none of which has been found in Antarctica. But the investigators have hopes that one day in the bleak reaches of the Horlick Mountains someone will find the fossil remains of the free-swimming reptile *Mesosaurus,* which has appeared in places as widely separated as Brazil and South Africa. Though no vertebrate land animal ever has been found in Antarctica, Long and Doumani have seen in the sedimentary rocks "impressions of trails and tracks on bedding planes," which lead them to believe that the fossil remains of small reptiles will turn up. If *Mesosaurus* actually lived in Antarctica, this evidence would certainly be another strong link to the reality of Gondwanaland.

The natural history of Pangaea

The Gondwanaland fossils in Antarctica fill in the picture Suess began to draw in the nineteenth century. Australia, India, Africa, and South America do not complete the image without Antarctica at its center. Antarctica is the keystone of the supercontinent in the south.

But what of the northern hemisphere? For Gondwanaland, the Wegenerians substitute the Austrian professor's "Pangaea." Let us quickly review the major evidence for this concept, which sounds medieval.[8]

Item: As noted earlier in this chapter, the Appalachians, which are "interrupted" by the North Atlantic Ocean, continue through England, Scotland, and Norway.

Item: Massive rock folds in Africa, which developed during the lower Triassic, about 160,000,000 years ago, appear to stop at the coastline. One picks them up again in South America.

Item: The coal measures of Illinois, Indiana, and Pennsylvania also run through England, Northern France, Germany, Poland, and the Soviet Union.

Item: Several families of worms and scorpions are found in South America, South Africa, India, and Eastern Australia.

Item: In New York State the Devonian, or Catskill, formation is strikingly similar to the Devonian, or Old Red, sandstones of England.

Item: Why are fossils of *Mesosaurus* (the Permian reptile not related to the giant reptiles that appeared later on) found in the widely separated sediments of Brazil and South Africa?

Item: A number of surveys show that land masses in the eastern and western hemispheres are slowly moving away from each other. From 1823 to 1933 Sabine Island, near Greenland, was surveyed four times. Its longitude shows a westward shift of about 1300 feet, or 13 feet a year. In West Greenland, three surveys at Godthaab between 1863 and 1922 indicate a shift to the west of 12 feet a year. According to the United States Coast and Geodetic Survey,[9] the gap between Washington, D. C., and Paris, France, is increasing at the rate of 1 inch a year.

Item: The geology of California shows it is possible for sizable pieces of continental rock to move. The San Andreas fault, which runs through central California and beneath San Francisco, is cutting the southwest side of California off from the east by sliding it to the northwest. By matching rocks on either side of the fault it appears that the southwestern block has moved northwestward a total of 100 miles.

Item: The prevalence of southern hemisphere pine is cited as an indication of linkage of the continents. So is the distribution of a southern hemisphere earthworm called *Acanthodrilid,* found in New Zealand, Australia, and South America. From a study of fresh-water crustaceous animals of Tasmania, it was concluded that certain species of this fauna reached their present range, as the worms may have, by means of an Antarctic connection between Australia, South America, and New Zealand.[10]

One can see from this partial recital of data which appear to support a linkage of land masses—in the southern hemisphere at least—that a supercontinent is quite a handy thing to have to explain a number of biological and geophysical puzzles. In a letter written in 1881 to the surgeon-botanist Hooker (the good doctor who had sailed into McMurdo Sound with Ross), Charles Darwin needed a polar extension of Gondwanaland to settle some matters in his own mind. He stated:

Nothing is more extraordinary in the history of the vegetable kingdom as it seems to me than the apparently very sudden or abrupt development of the higher plants. I have sometimes speculated whether there did not exist somewhere during the long ages an extremely isolated continent, perhaps near the South Pole.[11]

When Darwin inquired about the isolated continent, no one had yet set foot in Antarctica, and the fact that it existed could only be surmised. If Antarctica had once borne the missing links in the evolution of plants, the evidence lay well buried under eight million cubic miles of ice.

A. C. Seward, the English botanist who reviewed the findings of the *Terra Nova* expedition in 1914, felt as Darwin had:

Meager as it is, the material collected by the polar party [Scott's ill-fated dash to the pole] calls up a picture of an Antarctic land on which it is reasonable to believe there evolved the elements of a new flora that spread in diverging lines of a Paleozoic continent, the *disjuncta membra* of which have long been added to other land masses where are preserved both the relics of the southern flora and of that which had its birth in the north.

This idea fits neatly into the theories of Suess and Wegener. Wegener dates the break-up of Pangaea in Cretaceous times, which is quite recently, as geologists think—only 60,000,000 to 125,000,000 years ago. The Cretaceous is the longest and last of the periods of the Mesozoic Era. When it ended, the Cenozoic Era, the age of mammals and of man, began.

Wegenerian theory was amplified in the 1930s by the South African geologist Alexander L. Du Toit of Johannesburg. He insisted that the role of Antarctica in Gondwanaland was central—the keystone state of that union.[12] Around East Antarctica, he wrote, "with wonderful correspondence in outline, the remaining puzzle pieces of Gondwana can with remarkable precision be fitted." The angle of Princess Martha Land conforms to the indentation at Beira; Princess Ragnhild Land lies next to the southeastern corner of Madagascar. Enderby Land fits its opposite, the coasts of Ceylon and of Eastern India. The great bulge of Wilkes Land fits into the Australian Bight, while Tasmania and the submarine bank south of it fit into the Ross Sea. New Zealand, restored to its former place next to Australia, is brought into line with King Edward VII Land, while the Antarctic Archipelego (the Panhandle) connects via a South Antillean arc with Patagonia.

Wegener and Du Toit hypothesized that the separation of the Americas from Europe and Africa was a consequence of the upthrusting of the Andean fold—the Canadian Rockies, the American Rockies, the Andes, and the Antarctandes. Du Toit believed that the Andean fold continued through Antarctica into the Pacific Ocean and emerged in Australia.

Until 1957, the mountain chain that had been traced southward down the Panhandle seemed to fade out in Ellsworth Land. But, as we have seen, Bentley's second expedition in Marie Byrd Land (1957–1958) found evidence that the Sentinel Mountains appeared to be a continuation of the Antarctandes. Also, the Pirritt-Bradley traverse from Ellsworth to Byrd stations (1958–1959) discovered that this system continued under the ice toward the polar plateau.

Whether the Andean fold actually runs through Antarctica into the Pacific will be determined only after a great deal more investigation, especially on the polar plateau. As was mentioned earlier, the Andean Mountains of Antarctica are much younger geologically than the massive Trans-Antarctic chain along the border of Victoria Land, and the two mountain systems have nothing in common except that they are mostly submerged by the same ice cap. In the early years of Antarctic exploration, the Trans-Antarctics were called the Great Antarctic Horst, a term meaning uplifted blocks of rock. Similar horst regions—huge, uplifted blocks of stone between two fault zones where the crust appears to be fractured—were known in Africa and South America. Geologists with Scott's expeditions regarded the horst of South Victoria Land as typical of raised continental margins throughout the fragments of Gondwanaland.

Opponents of the supercontinent theory

Though the theory of Gondwanaland is a neat and convenient way to explain Antarctica, until recently its supporters were far outnumbered by physiographers who declined to take the idea seriously—even though they used the term "Gondwana" as though they did believe it.

One of the formidable opponents of the Gondwanaland theory was Griffith Taylor, the senior geologist of the *Terra Nova* expedition, who dismissed the whole idea of continental drift as unsound. "The similarities between the faunas and floras of the southern continents can be adequately explained without imagining that Africa, Australia, Antarctica, etc., lay cheek by jowl," he wrote.[13]

Some oceanographers tend to disagree with the continental-drift theory because they cannot imagine how the lighter, granitic continents can move through the heavier basalt of the ocean bottoms. This view has been expressed by Dr. W. Maurice Ewing of Columbia University's Lamont Geological Observatory. He contends that the sea-bottom structure is too rigid to allow continents to sail through it, and that the relative positions of continents and oceans have remained the same since their formation.[14] Ewing and an associate, Dr. William L. Donn, of the department of geology and meteorology at Brooklyn

College, propose that the entire crust of the earth rotated as a unit, like a bowling-ball cover around the ball. With this turning, the Arctic Ocean moved to the north pole of the earth's axis, which formerly had been in the Pacific Ocean, and Antarctica moved over the south pole, which formerly had been in the Indian Ocean. This is a part of the Ewing-Donn theory of climatic oscillations as the cause of intermittent ice ages in the northern hemisphere.

According to the Ewing-Donn theory, the shift in the earth's crust took place 1,000,000 years ago. It came at the dawn of the Pleistocene Epoch and ushered in a cycle of alternating ice ages and interglacial times. For all we know, the cycle is still going on. We may be living today in the interglacial phase of it, well along toward the onset of the next glaciation. If it is as extensive as the last one, the next ice cap to creep down from the north would bury our civilization down to the Ohio River in this hemisphere and would wipe out most of Europe. This matter, therefore, while of academic interest to us, is likely to be an intensely practical problem to our descendants.

As a result of such a shift in the crust, both the Arctic Ocean and the Antarctic Continent assumed a new relationship to the sun. The spin poles of the earth are cold because they receive the least solar radiation, and the amount of sunlight reflected back into space there is high. Thus, when the Arctic and Antarctic areas moved to the poles, these regions became cold. Instead of rain, snow fell in Antarctica.

The Arctic Ocean, for a while, remained open water. Surrounded by land masses, it was a source of precipitation, which now fell as snow. On the lands around the Arctic basin, Ewing and Donn say, glaciers built up in the highlands and grew larger and larger. Under the relentless pull of gravity, they flowed southward, down over North America, Europe, and parts of Siberia.

Ice begets ice. As it extended over the land, the glacier reflected ever more sunlight back to space. Colder and colder became the land. A new climate was born—an Arctic climate. As it snowed for centuries, for millennia, the sea level slowly fell, for more and more water was being locked up on land in the great ice sheets, and thus millions of cubic miles of ocean water were now transferred to the land. Ewing and Donn found evidence of this sea-level change in the sediments they dredged up from the floor of the Atlantic Ocean and of the Caribbean Sea.

At length the level of the two oceans fell below a rocky sill which then separated the Arctic from the warmer waters of the Atlantic

Ocean between Greenland and Norway. Deprived of this source of warmth, the Arctic Ocean froze. With a lid of ice on it, it was no longer a source of precipitation. With their nourishment cut off, the glaciers slowly dwindled, and the climate gradually warmed. More and more land appeared, to absorb the sun's radiation and convert it to heat.

The glaciers shrank back to the north, surviving mainly on the Greenland subcontinent and the northern islands in the Atlantic Ocean. In North America and Europe forests regrew and animals returned. The sea level rose as water drained off the land. Once more the Atlantic Ocean rolled over the rock sill to melt the ice cap on the Arctic Ocean. Once more the Arctic became open ocean, a source for the heavy snows. And, again, the glaciers grew and spread.

In this manner, Ewing and Donn account for four glacial and interglacial sequences in the northern hemisphere during the Pleistocene Epoch.

While all this was going on, what was happening in the southern hemisphere? It snowed in Antarctica, and there was nothing to stop it from snowing. Except for the ice-pack apron around Antarctica, the Antarctic Ocean did not freeze. It is too big, too well warmed by the exchange of heat from the waters north of the Antarctic convergence. There was no shut-off of precipitation.

The great ice cap built up in Antarctica. It overflowed the land masses into the sea. It literally flattened the earth's crust. It was thick, as one would expect it to be after a million years of snowfall.

While Ewing and Donn allow the continents to change position with respect to the sun, they insist that the relationship of the continents to one another has always been more or less the same since they were formed. Thus the two New York scientists account for the climatic shift in the northern hemisphere and in Antarctica by rotating the crust around the inner mantle. Antarctica is moved from a point in the Indian Ocean to the south pole.

A third idea, that the continents may have moved apart as the result of the earth's swelling up like a balloon, has been proposed by Bruce C. Heezen, a colleague of Ewing at the Lamont Observatory. Heezen based his theory on the evidence of a global mid-ocean rift, which at least in the floor of the North Atlantic is getting wider. This suggested to him that the earth's crust is expanding.

Like Ewing and Donn, Heezen believes that the continents have maintained their relative positions with respect to one another. Instead of the crust's rotating, as in the Ewing-Donn theory, Heezen has it inflating, so to speak.

Rivers of rock in the earth

Conventional geophysical theory held that the evolution of the planet involved a temperate transition from hot to cool. As the earth cooled, it congealed, contracted, and became rigid. Mountain chains were upthrust by the contraction of cooling, in the same way as folds are produced when the ends of a carpet are pushed together.

This conception began to fall into disarray with evidence that the continents have grown by accretion, that there were several periods of mountain-building, that the earth is considerably older than the idea suggested, and, finally, that it did not cool down like pudding on a window sill. Inside the earth radioactive elements, such as uranium, thorium, and potassium-40, continue to decay, generating large quantities of heat, which might be a source of energy that could move continents around in any theory of continental drift. The original objection to Wegener's theory had been the lack of a convincing energy source. But with advances in oceanography it became evident that the heat of radioactivity deep in the earth set up convection currents within the mantle, the primeval rock between the earth's crust and its core. These turgid currents of hot, plastic rock reach the surface of the earth in mid-ocean, forming great ridges, such as the one which splits the floor of the Atlantic. Where the currents, moving at the rate of an inch or so a year, descend into the earth's interior, they pull down the crust with them, to form deep trenches. A network of marine-bottom ridges covering more than 40,000 miles has been sounded all over the planet, under the seas.

According to J. Tuzo Wilson, professor of geophysics and director of the Institute of Earth Sciences at the University of Toronto, the existence of these currents is confirmed by island structures in both the Atlantic and Pacific Oceans. Their lateral movement is explained by seismic evidence of a plastic layer of rock called the "asthenosphere," which may be several hundred miles thick. Observing that the thermal convection currents move through the asthenosphere, Wilson concludes, "Here is a mechanism in harmony with physical theory and much geological and geophysical observation that provides a means for disrupting and moving continents."[15]

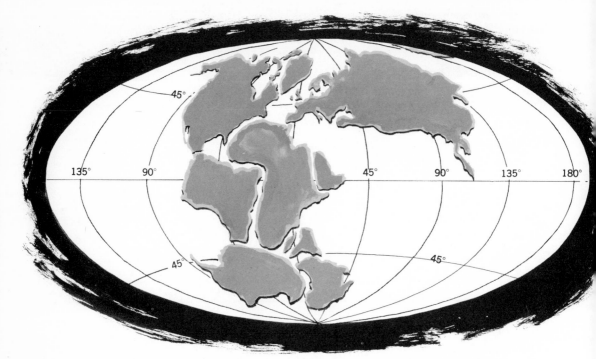

Reconstruction of continents during Mesozoic Era, assuming that continental drift was caused by spreading from mid-ocean ridges.

As the convection currents rise to the surface, the rocks are broken and thrust apart. In this way, Wilson visualizes, continental blocks were shoved sideways, "conveyed" (as on a conveyor belt) from the ridge, where the current rises, to the trench, where it falls back into the depths of the earth. This mechanism fulfills the drift-theory requirement of lighter continents plowing through the heavier basalt of the ocean floor.

Wilson points out that the age of islands in the Atlantic Ocean tends to increase with their distance from the mid-Atlantic ridge. This confirms the theory that many of them were formed by the upwelling of magma from the ridge, the older ones being pushed away as new material oozed out of the cleft.

On the basis of the ages of islands in the oceans and of continental coastal formations, Wilson proposes that as recently as 150,000,000 years ago all the present continents were welded into a single land mass. Thus, from his viewpoint, a Pangaea existed in Mesozoic times, surrounded by a planetary ocean, though this was not the first Pangaea. Wilson imagines periods of alternating assembly and disassembly of land masses—a continuous process, still going on.

The last supercontinent began to come apart at the start of the Cretaceous Period, about 120,000,000 years ago, Wilson estimates, when a rift opened in it to form the North Atlantic Ocean. In the south, South America pulled away from Antarctica, and Africa and India were moved away from the still joined Antarctica-Australia. About 60,000,000 years ago, another round of continent-tearing began. India was pushed away from Africa and, as it butted up against the bottom of Asia, the crust folded to produce the Himalayas.

About this time, Australia moved away from Antarctica, which presumably was jockeyed into its present position around the south pole. Then the ice came.

The United States Navy geologist Robert S. Dietz proposed a similar theory, in which the continents move by the spreading of the ocean floor from the outflow of rock through mid-ocean rifts. Dietz visualizes the continents afloat, like rafts on the denser sea floor.

Advocates of drift theory point to magnetic evidence that continental rocks have undergone significant movement in latitude and longitude. At the time they were molten, the rocks were magnetized sufficiently to align metallic particles with the earth's magnetic field. When the rocks cooled, this polarity became fixed so that where it is no longer aligned with the magnetic field it indicates that the rocks have moved.

Critics of the continental-drift theory take another view. Instead they suggest that perhaps the poles have wandered; perhaps the earth's spin axis has changed.

In any event, magnetic alignment in present rocks indicates that the north pole was in the Pacific Ocean at the time the rocks were formed.

Ice age upon ice age

The heart of any theory about Antarctica is the history of the ice cap. If it can be fathomed, such moods of the earth as climate and such processes as crustal movement may be clarified.

The age and evolution of the ice may tell us whether ice ages are products of terrestrial or solar changes, or both. But there have been other glaciations, in other times. We think of ice ages in the context of the last million years of geologic time, the Pleistocene Epoch. This is a scratch on the diamond of time. If the age of the earth is 4.5 billion years, the Pleistocene represents a small fraction of the history of the earth. What happened before that?

Glaciation has been suspected or proved in all the geologic ages in addition to the Pleistocene—the Eocene, Cretaceous, Jurassic, Triassic, Permian, Carboniferous, Devonian, Silurian, Ordovician, Cambrian, late Pre-Cambrian, Huronian, Sudburian or Timis-kamian.[17] The ice ages do not occur with regularity in the geologic record or in the same places. They appear to be nonrhythmical, haphazard in the grand scale of eonic time.

Why did India, Africa, Australia, and South America develop great ice sheets at the end of the Paleozoic, 200,000,000 years ago, when apparently Europe and North America were only lightly touched by the frost? During the Pleistocene, why were Europe and North America glaciated while the other continents, except for Antarctica, seemed to escape comparable ice sheets?

We live in a cool, dry period which may be merely a breathing space between glaciations. All about us, we can see the effects of the retreating ice age from which the northern hemisphere emerged only 11,000 years ago. In Sweden the ice began to leave the land about 13,500 years ago. Finland is still rising by isostatic readjustment to the removal of the load of ice. "Isostatic readjustment" is a rather cumbersome phrase for "springing back into shape." The weight of an ice cap literally bends the crustal rocks downward. When this weight is removed, by melting of the ice, the rocks return to their former position. This is a process that may take centuries, and in Finland it has been going on for a long time. In the nineteenth century, when Russia ruled Finland, it was said that enough land to form a new duchy arose out of the sea every eighty years.

Most geologists agree that the ice retreated out of Central Europe only 16,300 years ago.[18] In America, the length of post-glacial time has been estimated within a range of 7000 to 39,000 years. The most common date assigned to the end of the last ice age, the Wisconsin, is 11,000 years ago, an estimate determined by radiocarbon dating. Another measuring rod is Niagara Falls. Since the ice left that area of New York State, the falls have cut the gorge back 7 miles. It is estimated the cut took 20,000 to 35,000 years.[19]

We can arrive at approximate dates for the glaciations in this hemisphere because there are plenty of geological clues, but these are few and far between in Antarctica, where most of the evidence is buried. If we knew the age of the Antarctic cap, we would have one of the most important single facts about the earth's climatic history that mankind could find. Solving this problem is a primary objective of the Antarctic investigation.

But it is not a single problem. Tied inextricably to it is the history of the earth. Ewing, Donn, Heezen, and the "drifters" look to mechanisms within the earth for the answers. Advocates of synchronous ice ages caused by changes in the astrophysical environment of earth look to the sun. It may be that we shall find the answer on the moon, an ideal laboratory on which to study geophysical processes, or in further investigation of the sun. The answer to the Antarctic ice cap may lie in comprehending as a totality the environment of space.

On the other hand, it may be that changes in terrestrial climate are brought about by the interplay of events both within the earth itself and in its spatial environment. We see in Antarctica a double image. On one time scale, ranging over a half-billion years, it is forested at one end and glaciated at the other—a beautiful, shining hell of ice. And on a shorter time scale, covering a million years or so, we see Antarctica ice-covered, with some local variations in ice depth which suggest weak interglacials. If that is what they were, the interglacial periods were times when the ice was simply not quite so thick as at other times.

The change from the fair and open land to the ice cap, which shines bluely as it reflects the sky, is wonderful to think about. One looks over the brow of Erebus, the snowy, smoking volcano on Ross Island, over the frozen face of McMurdo Sound to the ice-clad Trans-Antarctic Mountains of Victoria Land. The land appears to be hibernating in the sun, as if one day it will awaken in the far future, to live again as in the far past.

We have only just begun to read the record of the ice cap. If we can probe far enough into the past, we might foretell the future.

Three photographers on Observation Hill set the camera to obtain a panoramic view of McMurdo Sound.

THE VAULTS OF TIME

On the morning of March 8, 1958, the icebreaker USS *Glacier* steamed out of McMurdo Sound into the Ross Sea, bound for Boston. Autumn storms had begun to sweep across the ice cap. Each day the sun dipped lower in the sky. The long polar night was not far off.

In her hold the *Glacier* carried a curious cargo—a ton of ice packaged in 5-foot-long cardboard tubes. There were 203 tubes, containing sections of a 4-inch core of ice, over 1000 feet long, that had been bored out of the West Antarctic ice sheet near Byrd Station by scientists of the United States Army Corps of Engineers' Snow, Ice and Permafrost Research Establishment, otherwise known as SIPRE.*

Here was the raw, unprocessed data for one of the most ambitious experiments of the IGY. Nearly one-fifth of a mile long, the core represented an estimated fourteen hundred years of snow accumulation in Marie Byrd Land. If a method could be found of determining seasonal variations in snowfall through the core, the SIPRE scientists would have an important new tool. With it they could ascertain a minimum age for the ice sheet. They could read a record of climatic history dating all the way down to rock bottom.

The great core was taken from the ice sheet with a conventional oil-well drilling rig between December 16, 1957, and January 26, 1958. The SIPRE crew laboriously sawed all but the bottom 160

* Now called Cold Regions Research and Engineering Laboratory.

146

The USS Glacier *carried the ice-core sections from McMurdo to Boston.*

feet of the core lengthwise, into half-rounds. At this point the last of the bandsaw blades at Byrd Station snapped, since the deep ice was as hard as a bar of steel.

By splitting the core, SIPRE was buying insurance. One half (plus the unsawed portion) was packaged in the tubes for shipment 10,000 miles to Wilmette, Illinois, a Chicago suburb. In this residential North Shore community SIPRE maintained a research laboratory in what had once been a steam laundry. There investigators could analyze the core in the "cold rooms" of the converted laundry's loft-like third floor.

The second half of the core was left in a shed at Byrd Station. If the first half melted en route to Wilmette, the SIPRE men could send for the second half the following year.

From Byrd Station, the tubed ice was flown 647 miles across the ice cap to Little America, then the main port of entry into Antarctica, where it was loaded aboard the *Glacier*.

The icebreaker steamed northward to Port Lyttleton, the harbor for Christchurch, New Zealand. From there, after her ice damage was repaired, the ship sailed across the Pacific Ocean and through the Panama Canal, and headed into the North Atlantic.

On the afternoon of April 13, 1958, she put into the Boston Navy Yard. The next morning the ice from Antarctica was to be unloaded into a refrigerator truck for the 1200-mile run to Wilmette. But it was a balmy spring day in Boston. The temperature was 62 degrees and it was raining. Could the ice be brought up from the *Glacier*'s refrigerators in the hold and transferred to the waiting vans without melting?

The problem was met in a practical way. The crew was organized into a bucket brigade. Hand to hand, they passed the cardboard tubes from the hold to the truck trailer, which had been cooled down with dry ice to supplement its refrigeration system.

Carrying its load of Antarctic ice, the truck bounced out of Boston and rolled southward through New England. It picked up the New Jersey and Pennsylvania turnpikes and crossed the mountains. Stopping only long enough to eat, sleep, and pay tolls, the drivers and a SIPRE representative kept moving across Ohio and Indiana and descended into Chicago from the Calumet Skyway Bridge. With its cold cargo, the truck made its way through Chicago traffic and rumbled into Wilmette. Earnest W. Marshall, one of the glaciologists who had supervised the drilling in Byrd Land, was there to receive the shipment.

Marshall and a SIPRE geologist, Anthony J. Gow, a New Zealander, went to work on the core analysis. The project of attempting to read the record of the past had begun. But this vault was not to be unlocked easily.

Sorge's law

The ice cap of Antarctica, theoretically, contains within it a history of climate since the first snow fell. It is therefore one possible key to the physiographic history of the southern ice age, and to a significant portion of the earth's past.

If scientists can learn how to turn this key, they may be able to discern the forces that change the climate and surface of our planet. But before they can reach these goals they must discover two secrets of the ice: how old it is and what is happening to it now—whether it is growing, waning, or in balance.

From the sediments on land and at the bottom of the seas, investigators have learned much about the earth's history, dating back 600,000,000 years, to Cambrian times. Records in the rocks reveal the ebb and flow of ancient seas, the rise and fall of landscapes, and the growth and decline of plant and animal life. With techniques of radioactive-isotope dating, the age of the rocks has been calculated

University of Wisconsin scientists use a power drill to bore through 15 feet of ice in order to lower instruments to the bottom of Lake Vanda.

over spans of millions of years by the decay of uranium and other long-lived isotopes. By a similar process, but over a much shorter time span, analysis of the carbon-14 isotope reveals the age of plants or animals that lived up to about 50,000 years ago.

The Antarctic ice cap cannot be made to give up its secrets in this way. Yet the length of time it has existed is critical to any theory of ice ages and climatic change. SIPRE's deep-core drilling program was the first attempt to open the great time vault of Antarctica and peer within.

To the glaciologist or glacial geologist, ice contains a complete record of the climate at the time it fell as snow. The record includes annual temperature, the amount of precipitation, and even the pre-vailing wind. All this can be read from an ice core, if one knows how.

It is also possible to determine the length of time it took the ice sheet to accumulate. That is not the same as the length of time a region has been glaciated, of course. Ice in West Antarctica is on the move, flowing down to the coasts, breaking off the main sheet, and floating away in the ocean as icebergs. It is probable that the ice we see there now may represent only a fraction of the time that Antarctica has been glaciated.

Theoretically, to determine the age of a section of ice, one would have to drill down to rock bottom and analyze the core all the way down. The SIPRE drillers had gone just about one-tenth of the way through the Byrd Land ice sheet. What they had obtained was merely a good-sized sample, but it was a beginning.

One of the earliest investigators to realize that a column of ice is a filing cabinet of climatic records was a Tyrolean glaciologist named Ernst G. Sorge. He was a member of Wegener's expedition to Green-land in 1930, and in many respects was as daring and imaginative in finding clues as the controversial innovator of the theory of conti-nental drift.

Near Eismitte in Greenland, Sorge studied the ice structure in pits 47 feet deep. Along the sides of pits he observed that the snow lay in alternating bands of dark and light hue. The layers were easy to distinguish, and Sorge saw in them a natural calendar. He knew that snow which falls in summer is wetter and therefore denser than winter snow. Also, summer snow in Greenland contained more wind-blown debris from the south. Therefore, summer snow was darker than winter snow.

The darker layers, then, consisted of snow which had fallen in the summer. All Sorge had to do to tell how long it had taken for the upper 47 feet of the ice cap to accumulate in the region near Eismitte was to count the dark bands, one for each year. Moreover, the dark and light bands were further distinguished by an intervening thin film of ice, which represented the refreezing of summer melt as the winter came.

Once these color distinctions were tied to the seasons, it became a simple matter to date a section of ice by counting the dark or the light bands. By measuring their thickness, the researcher could determine the annual snowfall.

Sorge showed that you could read an ice core's history in the same way as you would count tree rings. There was a relationship between snow density and climate.

Suppose one layer of winter snow should be less dense than another winter layer above or below. That indicated a change in winter climate. Sorge stated this principle in a law which said that, taking depth into account, the density of snow does not change with time unless the climate changes.

Sorge's law worked very well in Greenland, where summer-winter layering is quite definite and the refrozen layer of summer melt emphasizes the seasonal change. But the law was harder to apply in Antarctica, where melting does not occur in summer in many areas of the ice cap. Nevertheless, by applying Sorge's law to the Antarctic core, the SIPRE investigators hoped to read a climatic record of several thousands of years.

Dr. Roger Revelle, an oceanographer and member of the United States National Committee for the IGY, had explained the technique to a subcommittee of the House of Representatives Appropriations Committee at the time the Antarctic research budget was being considered in 1957:

"Now here in the glaciers you have a kind of library of what has happened in the past, locked up and frozen. It is a big icebox in which all of the geophysical events of the last million years are preserved.

"We may be able, by taking borings and studying the dust and the oxygen and nitrogen ratio and the carbon-dioxide content to get some idea of what has happened over the past. It is sort of a record."

Representative Albert Thomas of Texas, the subcommittee chairman, was puzzled:

"You are talking a little fast, there, doctor. You mean that you are going to measure the hydrogen and the oxygen and your other elements and that will give you some idea as to what has been taking place?"

"That is correct. For example, how many volcanic explosions have there been per decade or century or all over the past few thousand years? Here you have a layer of volcanic ash that will settle and form a distinct layer in the ice, and you can pick it up all beautifully preserved. . . . For example, with regard to the great explosion of Krakatoa in 1883, there is a distinct layer of volcanic ash from this explosion in Greenland. Another example—such things as changes in the ratio of gases in the atmosphere.

"Take a third example. We know—and the earth-satellite people are, in fact, very much interested in this—that the earth is constantly receiving very fine meteoric material, not only from the shooting stars but also from outer space.

"Here again you have this wonderful uncontaminated layer of snow and ice which is just a series of pages turning over. In them has been kept a continuous record of the amount of meteoric dust coming into the atmosphere."[1]

Dr. Revelle did not get a chance to explain Sorge's law. But the appropriation was approved so that the law could be applied in the Antarctic.

The core explodes

There were, however, limitations to its application in the south polar region. Below a depth of 400 feet the ice had become so compacted by pressure that the visual distinctions between summer and winter snow apparently were obliterated.

At Byrd Station only crude measurements had been possible. The core had been placed in a wooden trough in the station garage and viewed in the poor illumination of a 150-watt bulb. The thickness of the layers was measured with a tape, and sections were photographed. Though to the eye the ice seemed homogeneous below 400 feet in depth, the scientists hoped that they would be able to pick up the banding at greater depths in microscopes at the Wilmette laboratory. If that would not work, other means of reading the core would have to be devised.

During the drilling in Byrd Land the temperature had been measured in the drill hole every 50 feet. At the first 50-foot level it was 27.92 degrees below zero Fahrenheit. At 1000 feet, it had fallen by nearly 1 degree. If the core represented a 1400-year slice of Antarctic climatic history, it could be assumed that conditions were slightly colder in Marie Byrd Land in the sixth century than in 1957–1958.

A special problem had persuaded the drillers to halt the drill at 1013 feet. At this depth, gases trapped in the ice were highly compressed. When the weight of the overlying ice was removed, the gases expanded and shattered the cores. Unless a means was found to keep the cores under pressure, this depth seemed to be the practical limit of the deep-core-drilling approach to opening the Antarctic time vault.

The problem of drilling deeper than 1013 feet was shelved for the time being. Of first importance was a method of determining seasonal variations below 400 feet. Marshall and Gow applied themselves to this problem in the spring and summer of 1958.

Marshall settled on a technique he called "particulation," based on the observation that summer snow had more tiny debris in it than winter snow. In the Antarctic summer, the prevailing winds from the north blew in microscopic plant bits and rock dust from the tropics, whereas the big winter high-pressure systems tended to keep this debris out of Antarctica.

Marshall shaved off thin cross sections of the core with a razor blade, melted them, and peered at the water through a microscope. His assumption had been right. He found particles—all kinds, ranging from plant spores to volcanic, or possibly cosmic dust, as Dr. Revelle had described. Sections of the core which were relatively free of the debris were assumed to have accumulated in winter.

The counting of the annual layers was resumed, more slowly now, for the microscope technique was tedious. Marshall suspected that even the microscope was not revealing all the particles. Some were probably so small that they were beyond the instrument's resolution. He considered using an electron microscope but discarded the idea as it would stretch out the project to a lifetime.

A physician friend suggested that Marshall run the melt water through a blood-cell counter, an electronic device which reacts to blood cells passing through an electrical field and ticks off the total number instantly on a dial in values of 10 to 10,000,000 per cubic centimeter. This device could count other tiny particles just as well as blood cells.

Thus, SIPRE acquired a blood-cell counter. The first time that Marshall poured melt water from the core down its throat it clicked hysterically, like a Geiger counter in a field of pure pitchblende. As Marshall fed it melt water from the core, the machine's counting rate would rise and fall in alternating "pulses." Each pulse represented a cluster of microscopic debris in the ice. After a series of experiments, he was able to identify melt water from summer snow by the greater frequency of pulses, compared with melt water from winter snow. He compared the pulse frequencies in samples from the upper

400 feet of the ice core, and they checked with the visible banding: in the banded section of the core, the darker segments produced more pulses than the lighter ones.

Marshall's conception was something new. Potentially, here was an electronic method of measuring time in the files of the ancient ice sheet.

The scientists also attempted a qualitative approach to the debris in the ice. What did it consist of? How much was wind-borne organic drift from the tropics? How much was inorganic—rock dust, grit from volcanoes, or interplanetary dust falling from space?

There was an active volcano 450 miles from the drilling site— Mount Erebus on the west end of the Ross Sea. Today it merely smokes, but a century ago it was erupting. Dust from its eruptions must surely be in the ice. Possibly there was ash from the explosion of Krakatoa in the Sunda Strait in 1883, as Revelle had forecast. In Greenland SIPRE drillers had found volcanic ash from the 1912 eruption of Mount Katmai in Alaska. The ash served as a bench mark in time.

By qualitative analysis, Marshall believed it would be possible to identify the prevailing winds for fourteen centuries in West Antarctica. The clues might be grains of sand from the Australian desert, spores and other plant bits from South America or from Africa.

Reading the oxygen calendar

In the meantime, another means of determining summer and winter snow in the core, as well as climatic variations generally, was proposed at the California Institute of Technology. Two researchers, Samuel Epstein and Robert P. Sharp, both of the Institute's Division of Geological Sciences, suggested a system based on the ratio of normal oxygen, (O^{16}), to the isotope oxygen-18 (O^{18}).

With 18 neutrons and protons in its nucleus, O^{18} is heavier than ordinary oxygen, which has only 16 nucleons. Consequently, molecules of water formed of O^{18} tend to fall out of the atmosphere as precipitation before water molecules containing O^{16}. Hence H_2O^{18}, being less volatile than H_2O^{16}, precipitates at a higher temperature than water formed with normal oxygen. This has been tested a number of times by researchers, who found that tropical rain contains the highest concentration of water molecules with heavy-oxygen atoms and snow contains the lowest.

Applied to the problem of dating ice, a high ratio of O^{18} to O^{16}

would indicate that the snow fell in summer, and a low ratio in the snow would show it was winter snow. Preliminary tests on ice samples from Greenland and the Blue Glacier in Olympic National Park, Washington, proved that the oxygen-isotope method matched visible banding and thus presumably could be relied upon to identify seasonal snowfall below the level where banding was visible.

With two methods—particulation and oxygen-isotope—of discriminating between summer and winter snow in an ice core, it appeared to be worthwhile to attempt to get deeper cores for analysis, and this the Corps of Engineers proposed to do. The Office of the Chief of Engineers submitted a plan for using a new kind of drill on the ice cap—a thermal drill. Instead of a mechanical bore, the drill would be an electrically heated ring, which would melt its way through the ice. With this rig, melt water had to be pumped out of the hole, and the core passed through the center of the ring. A thermo-drill developed for SIPRE was taken to Greenland for testing before being shipped to Antarctica.

While the design of this new tool was taking shape, SIPRE resumed drilling in Antarctica in December, 1958. The mechanical oil-well drilling rig was moved from Byrd Station to Little America. The crew bored a hole almost through the Ross Ice Shelf, to a depth of 836 feet. The drillers did not dare perforate the shelf, for the waters of the Ross Sea lay underneath the great mass of ice at considerable pressure. Had the drill gone all the way through, the eruption of water from the hole might have destroyed the rig.

In the spring of 1959 tests of the thermo-drill in Greenland were started near Camp Century. Intermittent power failures were reported. SIPRE then decided to make modifications on the drill, which became increasingly complicated, and the prospect of boring all the way through 9400 feet of ice in Marie Byrd Land, an objective of the thermal-drilling proposal, began to fade.

At the end of 1960, SIPRE ran out of funds for analyzing the Byrd Land ice core. The 836-foot Ross Shelf core, which also had been brought to Wilmette, remained unanalyzed in the cold rooms of the research establishment. Marshall was unable to continue his particulation project. He left SIPRE in 1961 to join the faculty at the University of Michigan. Gow remained to study the cores, on his own, when he had time.

Then SIPRE was reorganized into a new kind of unit—the Cold Regions Research and Engineering Laboratory (CRREL), which was to emphasize applied research rather than basic investigations.

The great ice cores from Antarctica simply remained in cold storage. CRREL abandoned its quarters in Wilmette in 1962 and moved to Hanover, New Hampshire, taking the ice cores along with it.

Thus ended for the time being the initial project of coring the ice cap of Antarctica. Meanwhile, at Camp Century in Greenland, work was continued on the development of a 70-ton thermo-drill unit, requiring 60 kilowatts of electrical power, to be used in the austral summer of 1966–1967 for perforating the ice at Byrd Station down to rock bottom. This new project of the Office of Antarctic Programs has as its objective an ice core at least 8200 feet long with a diameter of 4.75 inches. About one-third of it, weighing 50 tons, would be shipped to Hanover for analysis.

Before leaving the Corps of Engineers, Marshall attempted to project his observations of the first 400 feet of the core to an assessment of the entire 1013 feet. His estimation that the full core represented about 1400 years of snowfall indicates that the 10,000-foot-deep ice sheet in Marie Byrd Land, from which the core was extracted, took from 14,000 to 20,000 years to accumulate.

This figure, however, is not the same as the age of the glaciation, which may be 10 to 100 times that old. The sheet of ice from which the core was taken may have replaced earlier generations of the ice sheet which have long ago flowed away.

Interglacials of McMurdo Sound

Can the massive Antarctic ice cap be a relic of some much greater ice age in the south, as the Greenland ice cap appears to be in the north? And if this is so, is it possible that the southern ice cap fluctuates with the same temporal rhythm as the ice ages of the northern hemisphere?

The possibility that oscillations of the Antarctic cap have corresponded with the northern glaciations of the Pleistocene Epoch has been suggested by the work of two investigators, Troy L. Péwé, staff geologist of the United States Geological Survey in Alaska and associate professor of geology at the University of Alaska, and Norman R. Rivard, a glacial geologist with the survey.

Between December 1, 1957, and February 10, 1958, Péwé and Rivard examined the McMurdo Sound region to see what its glacial history might have been. As the largest ice-free area in Antarctica, it held a particular fascination for them, for there was evidence that the region once had been under the ice. The area now free of deep

A University of California scientist measures the temperature of an Antarctic pond so salty that it never freezes.

ice, although it is covered with snow most of the year, ranges from 10 to 50 miles wide and is about 100 miles long. It extends along both shores of McMurdo Sound, including Ross Island on the east and the coastal mountains and valleys of Victoria Land on the west.

The recession of the continental ice cap from this region is plainly visible, not only from the striations, or scrape lines, on the bare rock walls of the valleys in Victoria Land but from their scoured look.

From the Victoria Land ice sheet a few big glaciers flow through the passes of the Royal Society Mountains into the Ross Sea. There are small Alpine-type glaciers in some of the valleys, but these are no longer connected to the continental ice sheet which overspreads Victoria Land to the west. The ice fronts of these glaciers look as though they had been cut with a giant saw. They are smooth and vertical, as though hand-hewn.

Some of these ice fronts were measured from 1910 to 1913 by Griffith Taylor, the Australian geologist with the Scott expedition, and again in 1957–1958 and 1958–1959 by American investigators based at McMurdo Sound. The positions of the fronts have not changed.

However, measurements of the position of the Ross Ice Shelf in 1899 indicated that it had receded 30 miles since Ross sighted it in 1841. The 1899 readings were taken by William Colbeck, a British Naval reservist who served as navigator and surveyor on Borchgrevink's expedition to the Ross Sea in the *Southern Cross*. The party wintered over at Cape Adare, and Colbeck made the first sledging trip on the Ross Shelf.[2]

Since 1899 there has been no similar recession of the Ross Shelf, but the findings of Péwé and Rivard are on another time scale that runs in millennia, not in decades.

On the walls of the mountain valleys in Victoria Land they discovered the telltale striations of high ice. And in the valleys lay stony debris, which the ice had pushed ahead of its advance and left piled up in heaps when it retreated.

The investigators reported that they could account for four separate advances and withdrawals of the ice cap in the McMurdo Sound region. All occurred during the Quaternary Period, in the last million years of geologic time, the period of the Pleistocene Epoch, when at least four major glaciations occurred in the northern hemisphere.

Each succeeding advance of the ice into McMurdo Sound was less extensive than the one before it. The oldest and biggest of the ice advances Péwé and Rivard called the McMurdo Glaciation. At its height, McMurdo Sound was choked with ice 3000 feet thick at maximum elevation. All that ice is gone today.

The next oldest advance of the ice is called the Taylor Glaciation. It filled the present dry valleys, such as the Taylor Valley, to an elevation of about 1000 feet. The striations of this ice are plainly visible on the rocky walls of the valley as horizontal scratches, scrapes, and gouges in the light-gray rock.

When the Taylor Glaciation receded it left several large lakes. One, called Glacial Lake Washburn, was 1000 feet deep. Its ancient shoreline still can be seen 1000 feet above sea level as a horizontal ridge in the valley wall.

The lakes bequeathed by the Taylor "ice age" vanished. Then the third advance, the so-called Fryxell Glaciation, appeared. It was less extensive than the Taylor, and it too left lakes, which also disappeared.

The most recent and smallest of the four advances of the ice is called the Koettlitz Glaciation. At its peak, the outlet glaciers extruding into McMurdo Sound were bigger than they are today, covering several islands in the sound to depths of 1000 and 2000 feet. The islands are now ice-free.

When did all this happen? The time sequence is rather vague. Algae swept up by the Koettlitz Glaciation were sent to the Lamont Geological Observatory of Columbia University for carbon-14 dating,

A skull-like rock formation, about 8 feet tall, sculpted by winds in a boulder field covering many miles in an ice-free region of Victoria Land.

which indicated that the fossil plants were from 2500 to 6000 years old. So it is possible that the Koettlitz advance began as late as 2500 years ago.

Since radiocarbon dating is not reliable for organic materials older than 50,000 years, it could not be accurately applied to the earlier glaciations. Péwé reported that he was uncertain whether the glacial sequences covered all, most, or just a part of the Pleistocene Epoch. He estimated that the most recent advance may have coincided with the last of the northern hemisphere glaciations, the Wisconsin. However, if one relies on the radiocarbon dates for the Koettlitz algae, this estimate may be stretching matters a bit. The Wisconsin ice sheet was gone 11,000 years ago, while radiocarbon dating indicates that the Koettlitz is much younger.

There is, it appears, only a suggestion here that the oscillations of the ice in McMurdo Sound compare with the northern hemisphere glaciations. A major question is, of course, whether the McMurdo Sound glaciations represent oscillations of the entire ice cap. They may be only local events, and consequently may have no connection with widespread climatic change. The task ahead for Antarctic investigators is to determine the answer to this.

However, Péwé and Rivard are certain of one thing. They believe the present ice retreat in the McMurdo Sound region represents a long-term trend, despite the advances and retreats of the ice, and that the trend is still continuing. This belief is supported by other scientists, who are convinced that, while the cap appears presently to be in equilibrium, it is waning on a rather long time scale.

During their probing of the dry valleys, the two Alaskan scientists found ninety mummified carcasses of crabeater seals, some thirty miles from the present seashore. Lying on valley slopes one to two thousand feet above sea level, the carcasses were well preserved in the dry Antarctic refrigerator. They were a puzzling discovery. How did the seals manage to get there?

At first the remains were estimated to be only 100 years old, but radiocarbon analysis of a piece of one carcass showed that the animal had lived 1600 to 2600 years ago. This particular carcass was found 1640 feet above sea level in glacial drift rocks on the slopes of Mount Nussbaum in the Taylor Dry Valley.

The remains can be explained in this way. The ice was high when the seals had wandered inland. Then, finding no food in the fresh-water lakes, the seals had perished. Subsequently the ice retreated, leaving the carcasses on the rocks.

On the basis of this evidence, there was high ice in the Taylor Valley between 1600 and 2600 years ago. That does not seem to correspond with any major climatic change in the northern hemisphere.

Rising beaches

Another Antarctic investigator is convinced that Antarctica is warming up and that the ice cap is shrinking. He is Dr. Nichols, the head of the geology department at Tufts University, who was mentioned earlier as one of the first Antarctic scientists to predict a thick ice cap.

Nichols has studied other clues to ice retreat in the McMurdo Sound region—rising beaches. Along the coast of Victoria Land the beaches are actually rising in elevation, and so are beaches on the coast of the Panhandle, 1000 miles away. Directly across McMurdo Sound from the Naval Air Facility on Ross Island there are a series of raised marine beaches in front of the Wilson-Piedmont Glacier. They show that the glacier has been retreating from the coast.

When the ice of the glacier extended farther seaward than it does today, these beaches were pressed down, like all the ground underlying the ice. Then, since the ice shrank back, the beaches have been slowly rising as an isostatic readjustment to the removal of the ice load.

In the Panhandle, Nichols found that the ice at one time had extended into deep water. Today it is well back from the shoreline. The beaches rose 100 feet in most places. Nichols estimated that the rise represents a former ice load of 300 to 400 feet.

On the west shore of McMurdo Sound, a lucky accident made it possible to ascertain the rate at which some of the beaches have been rising. It was the discovery of the carcass of an elephant seal by a bulldozer which was smoothing an emergency air strip. The seal was lying in gravel 44 feet above sea level, and radiocarbon dating of a sample of the carcass showed that the animal had lived about 4600 years ago. Assuming that the beach had been at sea level then, Nichols concluded that it had risen 44 feet in that time—a little less than a foot a century.

The glowing rocks

One of the confusing aspects of attempting to find a trend in the ice cap is the question of time scale. Scattered all over the continental ice sheet are bits and pieces of evidence that there once was more ice than now. But when?

In discussing Gondwanaland, scientists are thinking of a time scale calibrated in tens of millions of years. The interglacials of Péwé and Rivard may be figured in tens of thousands of years. And the efforts to analyze the ice cores represent a time span of merely thousands of years.

The problem here is one of perspective. And we cannot judge temporal distance in this timeless region until some means is found of telling when the present glacial period began.

The question that must be asked, then, takes this form: how long has Antarctica been continuously cold—cold enough to have caused the present ice cap to form?

An ingenious method of trying to answer this question was devised by Dr. Edward J. Zeller and his associates at the University of Kansas. It was based on a means of finding out how long calcitic rocks, such as limestones and marble, have been continuously at freezing temperatures. It is possible, theoretically, to determine this by measuring a characteristic of these rocks called thermoluminescence. Such rocks that have been cold for a long time shine when heated. The luminescence may not be visible to the unaided eye, but it can be detected and measured very accurately with the aid of a sensitive device called the microphotometer.

Over eons of time, energy is accumulated in the rock's crystalline structure by the decay of such radioactive elements as uranium or thorium within the rock. At low temperatures the energy remains in storage—cold storage, so to speak. But when the rocks are heated, the stored energy is released in the form of light.

With such a project in mind, Zeller began prospecting for calcitic rocks in the McMurdo Sound region during the austral summer of 1958–1959. He chipped off samples of marble at Marble Point on the coast of Victoria Land. He took limestone samples near the Taylor Glacier.

Having packed the rocks in dry ice to keep them cold, he hauled them back by airplane to his laboratory in Kansas, where he ground the rock into powder and heated it at a rate of 10 degrees per second. As the rock dust became warm, it emitted light in bursts, or pulses, at several different temperatures.

There was one luminescent display at 122 degrees Fahrenheit. Then it died out. Another flash appeared at 257 degrees, a third at 392 degrees, and a fourth at 572 degrees.

Zeller had found in testing temperate-climate rocks that they did not emit any light at 257 degrees Fahrenheit or below. The energy they had stored up had been triggered long ago by the temperature of the environment.

Only those rocks which had been cold for long periods of time gave off light at 257 degrees. Zeller estimated that at this temperature the rocks emitted the energy they would release naturally when the climate warmed above freezing.

The storage of energy is caused by the displacement of electrons in the crystal lattice of the rock as a result of the bombardment by alpha particles from the decay of radioactive elements within the rock. The alpha particles go whizzing around, like a cue ball on a billiard table, knocking the billiard-ball electrons out of their normal positions in the lattice. Some of the electrons fall back into place, but others don't. They fall into other "pockets"—holes in the lattice structure, which are energy-level vacancies, due to electronic imperfections in the crystalline rock. The displaced electrons remain in the "pockets" until some application of energy, such as heat, causes them to leap out, like Mexican jumping beans. When they do, they lose the energy they gained when hit by the alpha particle, thus balancing the books. The energy is given up in the form of multiples of the photon, the unit of light.

These photons comprise the glow that the microphotometer sees and records when the rocks are heated. The deeper the electron is

Near Walcott Glacier in Victoria Land, a University of Wisconsin geologist inspects patterned ground as a means of dating the advances and retreats of the ice sheet during the past several thousand years.

enmeshed in the ion trap, the more heat is needed to flush it out.

Once Zeller had found that only rocks which had been at freezing temperature or below for a long time would emit a glow at 257 degrees Fahrenheit, he had made a beginning. The next step was to find out how long the rocks had been that cold. One way to solve the problem was to determine how long it took the alpha-particle barrage to produce the energy in the glow.

First he put back the luminescent energy in the rock samples by subjecting them to high-energy radiation emitted by radioactive cobalt-60, which displaced the electrons just as the alpha particles had. It took a dose of 1000 roentgens to restore the energy of the glow that the rocks had emitted at 257°F. Zeller then estimated that this dose was equal to the natural bombardment by alpha particles at the rate of 1 particle per hour for 2,500,000 years. The natural bombardment rate from the radioactive elements in the rocks, however, had been found to be 14.5 particles an hour. Dividing by this rate, Zeller calculated that the limestone had been at freezing temperature or below for about 178,500 years. At least, that is how long it appeared to have taken the luminescence to accumulate in the rock at the 257-degree release level.

But could this kind of measurement be taken as an index of climate? Perhaps the sun had warmed the rocks at that time, 178,500 years ago, without any climatic changes necessarily having occurred.

Zeller was not satisfied with his results, although they looked promising, as some of his limestone samples had not emitted any glow at all at 257 degrees. There was nothing to do but to go back and get more specimens—which he did.

This time he extracted rocks from pits dug in places which had been covered until fairly recent times by the Wilson-Piedmont Glacier, thus reducing the possibility that some of the samples might have been heated by exposure to direct sunlight. These he carefully packed in dry ice and flew back to Kansas.

From this second batch of limestones he obtained better results. Some of them showed thermoluminescent values indicating they had been continuously cold for 220,000 years.

But, as he analyzed the samples, Zeller noticed a curious discrepancy. The rocks nearer the surface of the ground had been cold longer than those deeper down. Now what could that mean?

He checked and double checked his procedures. He was baffled. Then it occurred to him to take a hard look at the map. There lay the answer.

In a bright, luminescent flash it dawned on him that less than fifty miles away from the site of the rocks sat the enormous smoking volcano, Mount Erebus. The deeper rocks had been warmed by volcanic heat. In attempting to find out how long the region had been cold, Zeller had also hit upon a means of determining the diffusion of volcanic heat through the upper crust of the earth.

It was a wry bit of serendipity (an unexpected, but fruitful, discovery), and it began to turn some mental wheels. Could this be an explanation for the retreat of the ice from the McMurdo Sound region? Perhaps the climate hadn't changed. Maybe it was just old Erebus, waxing and waning with the pulsations of heat deep within the earth's crust.

The time values obtained by Zeller did little to support a theory that warming had caused the interglacials for which Péwé and Rivard had found evidence. If interglacials had occurred in the McMurdo Sound region, they were cold ones, and more likely the result of variations in snowfall than of changes in temperature. However, so far as his basic question was concerned—how long the region had been continuously cold—Zeller for all practical purposes had been foiled again.

In November 1960 he returned to McMurdo Sound with an assistant, Luciana Ronca, an Italian geophysicist. This time they dug underneath the Wilson-Piedmont Glacier. In addition, field parties working their way through the Trans-Antarctic Mountains that summer brought calcitic rocks back for the thermoluminescent study. By the end of the season Zeller and Ronca had rocks from a wide area of Antarctica, ranging from McMurdo Sound to the Beardmore Glacier and eastward along the rim of the polar plateau to the Horlick Mountains.

Now higher thermoluminescent values appeared. Some of the rocks gave glow values indicating they had been continuously cold for 350,000 years. There is thus some evidence that the Antarctic ice sheet has existed more than one-third of a million years. That is a beginning, but only a small one.

On the west shore of McMurdo Sound is a place called Gneiss Point, where the bedrock was shown by radioisotope analysis to be composed of sediments laid down 520,000,000 years ago. Ultimately the investigation of Antarctica may present the outlines of its climatic history back to the beginning of the Cambrian Period, half a billion years ago. And yet, in the scale of time the earth has existed, even that is relatively "recent."

*An accumulation of thick ice decorates the icebreaker
USS* Edisto *after she battled storms for six days in Antarctica seas during the winter season, in an attempt to
recover an oil tanker broken loose from her moorings.*

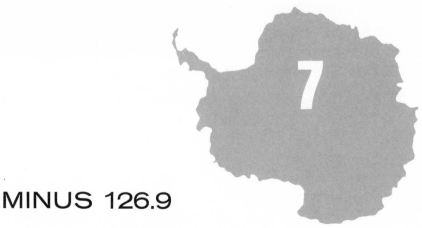

MINUS 126.9

DEGREES FAHRENHEIT

Climatically speaking, we live in a most unusual period of the earth's history. Since the Pleistocene Epoch, cooler temperatures have prevailed than during most of geologic time. In the last million years there were at least four ice ages in the northern hemisphere, but even in this cool million-year period intervals as long as 200,000 years appear to have been warmer than it is now, judging from the records of the rocks and their fossils.

Some 35,000,000 to 40,000,000 years ago our planet was considerably warmer than at present, and there were no polar ice caps. This kind of planetary climate, with only a small differential in average temperature between the equator and the poles, is considered by some investigators to have been "normal" for the earth, and our present climate, which supports ice caps on Greenland and on Antarctica, is considered distinctly abnormal.

According to one estimate,[1] since the beginning of the Cambrian Period, some 600,000,000 years ago, the earth's climate in the middle and higher latitudes was warmer than now about two-thirds of the time. Equatorial climate during the warm periods was the same as now.

Only 15,000 years ago the climate was cool enough in the northern hemisphere to support an ice sheet over the northern one-third of North America and Eurasia. Since then, though the temperate lati-

167

tudes have a more benevolent climate, it has remained cool enough
to support large polar glaciers.

Man has expanded over the habitable regions of the planet and
has evolved a series of cultures of ever-increasing complexity since
the last retreat of the ice. His number is now increasing at an
astonishing rate. As the population grows, mankind's dependence on
a stable climate becomes more critical. Any changes which affect
supplies of food, water, and living space are likely to be serious
threats not only to his comfort but to his physical survival.

In any consideration of climate, one is always brought face to face
with the prospect for the future. At the present level of scientific technology, we cannot control or predict climate change. We seem to be
too preoccupied by the threat of nuclear war to worry about possible,
or even probable, climatic calamities. Yet it is obvious that, if our
population continues to grow, a climatic change leading to another
glaciation of the northern hemisphere would be catastrophic—at least
for human civilization as we know it.

Where are we headed? Present meteorological evidence is being
rather widely interpreted to show that our climate in North America
is warming up. The trend began about 1850 and is likely to continue
for at least 200 years more. While even a short-range change may
affect us adversely in some areas agriculturally, we can adjust to it.
But the larger the population grows, the harder the adjustment will be.

What changes the climate? What can we do about it? These questions are not simply academic ones. They have meaning in terms of
the survival of our descendants. They must be answered if men are
to become masters rather than pawns of continuous environmental
change. It is essential to know whether the mechanics of climatic
change reside in the earth, as Dr. Ewing has suggested, or in changes
taking place in the sun or in the earth-sun relationship.

In Antarctica the southern ice age presents a natural laboratory
for the investigation of this problem. Perhaps here the scientist will
discover how an ice age begins, how it grows, and how it ultimately
dies.

To the glacial geologist, the problem of whether the ice cap is
growing, receding, or standing still requires the analysis of the ice
itself. The geologist tries to read the files—the petrologic record, as
though ice were rock, which it is to the geologist.

But the meteorologist takes another tack. For the weather scientist,
the quest involves relationships between temperature, atmosphere,
and ocean.

Ice is water which is evaporated out of the ocean and precipitated on the land through the atmosphere by the energy of the sun. Heat is one of the factors, and moisture is the other.

Scott observed in 1902 that a warming of the climate in Antarctica would not reduce the ice sheet but enlarge it. A slight warming would increase the air's capacity to hold moisture and thus increase the snowfall. Consequently, a warming in Antarctica might actually cause the ice cap to expand, while a general cooling, bringing drier conditions with it, would cause it to recede. For sustenance, a glacier requires a balance between temperature and moisture.

The precise nature of the balance is a problem the meteorologist may be able to solve. If he can figure it out, and devise a means of upsetting this balance, he may learn how to suppress the growth of a glacier, just as hail-suppression is being attempted today.

The annual deposit of snow on the Antarctic ice cap has been estimated at 4.25 to 6.54 inches.[2] In water equivalent, this amounts to all the water held at any one time in the atmosphere of the entire planet.

If the ice cap is in balance, as some experts believe it is today, it means that as much water is returning to the oceans through melting and the breaking off of icebergs as is being deposited on the cap. The best evidence that the cap is in balance is a constant sea level. It appears to have remained stable for at least the last half-century, and probably a great deal longer.

Rapid melting of the ice, of course, would cause the sea level to rise. The estimate which Antarctic scientists seem to agree on is that if the Antarctic cap should melt, the rise in sea level would amount to 240 feet. That would be as great a catastrophe as another ice age, for it would inundate many of the earth's major cities. If we look at the 240-foot contour line, we find that this much rise in world sea level would deprive us of most of the coastal cities of the western hemisphere from Boston to Buenos Aires and most of the ports of Europe. The District of Columbia would be sea bottom, with the top of the Washington Monument protruding above the waves, and the Atlantic Ocean would be surfing against the foothills of Appalachia. A new inland sea might roll in the Mississippi Valley, like the one that rolled there in Cretaceous times, 100,000,000 years ago.

Since we know there were times in the past when the Antarctic ice cap did not exist, we know it is possible for such an inundation to happen. What we don't know is how or when.

It is fortunate, then, that we seem to be living in a cool period.

It keeps the frozen sea on Antarctica frozen. Yet it is not cold enough or zonal enough to bring a new wave of ice cascading down upon us in the northern hemisphere. How stable this regime is remains to be seen. It has lasted some 11,000 years, but on the face of the great clock of geologic time, that is merely a second.

There is only one certainty: the climate will change, as it always has.

Weather Central

Until weather observations could be made all over the continent, there was no way of knowing what the Antarctic weather pattern was, how it affected the ice cap, or what its influence was on the southern hemisphere. There were a number of guesses, however.

John Murray, the Canadian scientist who had forecast the existence of an Antarctic continent in 1893, had predicted also that a zone of permanent high pressure lay above it. He was nearly right, so far as the high plateau of East Antarctica is concerned.

Later, mysterious "pressure surges" were thought to originate in West Antarctica. They were hypothesized by George C. Simpson, chief meteorologist with Scott's *Terra Nova* (1910–1913) expedition.

This picture of a static high-pressure zone in the east and "surges" in the west characterized the meteorological profile of Antarctica up to the IGY. In 1956 two of America's leading authorities on Antarctic weather, the late Harry Wexler and a research associate at the United States Weather Bureau, Morton J. Rubin, commented, "That Antarctic regions are cold we have known ever since Captain Cook's voyages of 1772–1775, but how cold they are we still do not know today."

With the mass invasion by science at the start of the IGY, it became possible to have synoptic, or simultaneous, weather observations all over the ice cap. The problem of organizing the data was settled by the creation of an international organization called Antarctic Weather Central. Its function was to collect data radioed to it from the stations on the ice and serve as a clearing house for the raw reports. In addition, Weather Central meteorologists were to make analyses and broadcast forecasts for Antarctic oversnow traverses and flying operations.

The special international committee coordinating scientific work during the IGY asked the United States to operate Weather Central.

Left: *An automatic weather station being installed at the foot of Beardmore Glacier.*
Right: *Power from the windmill operates devices that record ground temperature.*
Below: *An aerologist taking sunrise meteorological readings at McMurdo Station.*

The United States National Committee agreed and set it up at Little America V. The center was staffed with an international assortment of meteorologists from Argentina, Australia, France, New Zealand, South Africa, and the Soviet Union, in addition to American personnel. The first wintering-over group consisted of six meteorologists, four American, one Russian, and one Argentine. In subsequent seasons Australian, French, and South African experts were added to the staff.

American weather men were given a two-month course in southern hemisphere analyses in Washington before they departed for the southern continent in 1956, and reached Little America at the end of December. The experience was described by W. B. Moreland of the United States Weather Bureau: "At this time the camp was overcrowded with U. S. Navy construction personnel who were erecting buildings and unloading supplies from cargo vessels. The space allotted to Weather Central was being used as an operations headquarters as well as for housing the amateur radio equipment."[3]

During the first six weeks the Weather Central meteorologists spent

Personnel sorting out luggage upon arrival during a severe spring storm.

most of their time loading and unloading supplies from cargo sleds and "performing many other household duties necessary in establishing a station in this remote region of the earth," Moreland related.

Weather Central depended on an elaborate network, all over the ice cap, of so-called mother-daughter stations which had been devised by an IGY radio group under the chairmanship of A. H. Sheffield of England. Its operation depended on a highly integrated station-relay system. The mother station would collect weather reports from several daughter stations on a prearranged schedule. It would then relay these reports to a central collecting point at McMurdo Sound, where they were passed on to Little America.

Since very little ever happens in the Antarctic on any schedule devised by man, it was not surprising that Weather Central did not get under way until mid-1957. The Naval Air Facility antenna, which was blown over by a blizzard in May, was not back in operation until late in June.

Broadcast schedules were set up to come as close as possible to midnight, 6 a.m., noon, and 6 p.m., Greenwich Mean Time. The system worked fairly well. Belgian and Norwegian stations in the Australian quadrant radioed observations to a mother station at Australia's Mawson base, which relayed them to McMurdo. Mirnyi served as the mother station for a cluster of Soviet camps.

At McMurdo radioteletype circuits were opened between New Zealand and Little America. And, at Little America, Weather Central scientists not only sorted the enormous sheaf of Antarctic data they received four times a day but also monitored weather reports from points throughout the southern hemisphere. On June 3, 1957, the first weather maps of Antarctica were transmitted by Little America by radio facsimile directly to the United States Weather Bureau in Washington.

Weather Central remained at Little America until 1959, when closing of the station required transfer of the international facility to Melbourne, Australia.

The effect of Weather Central in improving the safety and efficiency of Antarctic transportation can hardly be overestimated. I took part in a resupply flight to the south pole from McMurdo Sound on November 14, 1959. It depended entirely on the accuracy of a Weather Central forecast and some additional data from the reporting network.

Several hours before take-off the air was filled with light, blowing snow. The weather deteriorated rapidly, and a partial polar whiteout enveloped the McMurdo Sound region. A whiteout is an optical condition wherein the light is reflected between the snow surface and a ceiling of clouds. All contrast disappears, the snowscape and sky seem to merge, the horizon vanishes, and it becomes difficult to determine relative distances. Moving in a whiteout becomes dangerous because it is virtually impossible to see a crevasse or a hole until you are right on top of it, so lacking is the contrast of normal light and shadow.

It appeared as though the mission would be scrubbed. Then Navy staff meteorologist William S. Lanterman told the Air Force crew flying the mission it could go. Those of us whose flying in the region had been limited to the flight from New Zealand were amazed.

The whiteout seemed to have lifted partially, but the air was still thick with blowing snow and mist as our party rolled down the volcanic beach of Ross Island in a Weasel (tracked jeep) caravan. We headed out across the frozen sound toward Williams Field, the airport on the bay ice about four miles from the island. Atmospheric conditions were so bad I could hardly see the vehicles ahead of or behind us. The lead Weasel was "piloted" by Lloyd M. Bertoglio, then commander of the base.

With a confidence born of experience, Bertoglio drove his Weasel, called the *Mayor of McMurdo,* into the blinding whiteness at what seemed to me to be a reckless speed, though it was probably no more than 15 miles an hour. He was following a trail marked by orange pennons on bamboo poles stuck in the ice. Beyond the flags, the ice was considered unsafe. The spring was far enough advanced at the time so that weak spots had developed in the thinning ice of the sound.

Eventually the semi-tubular Jamesway huts of Williams Field—the only airport in the world built on floating ice—appeared through the mist, and I was relieved to see the familiar sign, leaning against the hut marked OPERATIONS, that read: WELCOME TO MCMURDO SOUND, PARADISE OF THE ANTARCTIC.

I could hear the engines of the C-124 Globemaster in which we were to fly to the pole warming up long before I could see the aircraft —and the C-124 was then one of the biggest airplanes in existence. It is a big freighter, slow, shaky, but dependable. In fact, its nickname is "Big Shaky."

The Air Force crew, directed by Captain Ted Bishop, had loaded the airplane with 31,000 pounds of lumber, plumbing, and other supplies for Amundsen-Scott Station. These were to be dropped by parachute on the polar plateau.

With this drop and another, scheduled five hours after it was completed, the Military Air Transport Service of the Air Force would wind

The author entering the Operations hut at Williams Field.

A caravan of tracked jeeps en route to the airfield.

up its delivery of 1600 tons of cargo to the pole during Operation Deep Freeze '60 (1959–1960).

In the hold of the Globemaster the wooden crates were secured by a network of metal straps. For delivery they would have to be positioned by the crew on rails over the drop doors.

We strapped in on canvas bucket seats, and without delay Bishop began his run-up, the rubber-tired wheels skidding momentarily until they took hold of the ice runway. He lifted the plane, weighing 195,000 pounds with its cargo, smoothly off the ice into the blinding veil of mist and snow. The flying warehouse lumbered southward along the edge of the Ross Ice Shelf into an unearthly dimension where ice and clouds blended into an infinity of off-whiteness.

Presently the earth began to reappear. To the west, I caught sight of some huge peaks, which we later identified as the 10,000-foot summits of the Britannia Range. Beyond it lay the white-linen ice sheet of Victoria Land.

The weather cleared very rapidly. It was as though someone had wiped fog off the windows. Ahead, the Trans-Antarctic Mountains rimming the polar plateau became sharply drawn against the blue-green sky. The sun was a dazzling glory to our right.

Now the polar approaches lay ahead. Bishop kept the Globemaster at 11,000 feet. The cabin is unpressurized, and even at that altitude we had to sniff at supplemental oxygen.

In the brilliant afternoon sunlight we sighted the titanic Beardmore Glacier, up which Scott and his companions had struggled in the futile race with Amundsen a half-century before. This enormous river of ice flowed in a lazy S curve 100 miles long and 30 miles wide at its junction with the ice shelf between the Queen Maud and Queen Alexandra Mountains.

Bishop climbed to clear the higher peaks, and we took more frequent sniffs of oxygen. Ahead, the polar plateau opened up, a featureless desert of snow extending into infinity.

I was riding up in the forward crew compartment of the aircraft when Joseph Siniuk, the navigator, who was looking through his pole sight, announced that we were "coming up on" Amundsen-Scott Station.

A thin dark line appeared on the white plain below and then resolved itself into the roofs of huts half buried in the snow. As Bishop began to turn, I could see a ring of oil drums which had been set in a circle around the point of the geographic south pole. The mosquelike radome stood on stilts above the snow surface.

A blowing snowstorm above the camp site at the South Pole Station.

"Here we go around the world in eighty seconds," said Bishop as he circled for the first drop. It was no exaggeration. The Globemaster was turning in a steep bank through all 360 degrees of longitude.

Inside the cargo hold, dropmaster Sidney Henderson and his crew shifted the wooden crates to the drop doors with an electrically operated overhead crane. At an intercom signal from Siniuk, they opened the drop doors.

Polar air typhooned into the warm hold. The icy plain, corrugated with sastrugi, rushed past below.

The dropmaster swung his upraised arm down in a chopping motion, and his crewmen, bundled in parkas and fur hoods, cut the lashings with knives they held ready. Crate after crate tumbled through the hatchway, parachutes billowing out behind. I believe we were 2000 feet above the ice at this time. One of the chutes failed to open, and I saw the crate hit the ice, bounce, and explode into pieces.

It took four passes around the station to complete the drop. A tiny red tractor below was crawling out toward the merchandise, which was strewn over several miles of ice.

Bishop roared the C-124 once more over the station, and those on the aircraft watching through field glasses saw waving figures in parkas on the ice below. Then the Globemaster headed back northward. In a few seconds the outpost became a hyphen on the blank page of the polar plateau, and then it was gone. The mountains loomed up ahead.

In the sunlit evening, we were once more flying over the great ice shelf.

We arrived at McMurdo Sound later that evening, and the air was clear as crystal. The sun was still bright to the west and the sky was a clean, pale blue.

"This was the forecast," Bishop said as we walked to the Operations hut after landing. "They told us it would be clear when we returned, and that's all we needed. They didn't miss."

With Antarctic Weather Central, flying in the Antarctic has become a rational operation.

Blackout

One of the important areas not covered by the Weather Central network of reporting stations is the vast one of the Pacific Ocean between Australia and the Antarctic coast of Wilkes Land. On December 9, 1960, the answer to that problem was suggested in the form of a photograph from the weather satellite *Tiros II*.

A description of the photograph showing weather systems over an ocean area 3600 miles long by 900 miles wide was radioed to Captain Lanterman from Washington. The picture was a freak, insofar as *Tiros II* was not in polar orbit. But Lanterman and other meteorologists saw in it a demonstration that a weather satellite in polar orbit would provide a substantial boost in data for Antarctic and southern-hemisphere forecasting.

Several weeks earlier there had been a dramatic demonstration of the weak link in Weather Central operations so far as forecasting was concerned. A few days after a solar flare was observed to erupt on the sun, the ionosphere, which normally reflects the radio waves and makes long-distance radio communications possible, was badly disrupted. In the southern hemisphere there was a partial blackout of radio communications, and a total blackout occurred above the Antarctic Circle, starting November 9.

I was in New Zealand at the time, awaiting air transportation to McMurdo Sound. When I arrived at Christchurch with a party of scientists on a flight from California, I was advised there were no communications with any station in Antarctica. Until this situation was repaired, there was nothing to do but wait in Christchurch.

The Navy had a Constellation standing by at Harewood International Airport, but would not let it fly the 1800 miles to McMurdo Sound without a weather advisory from Antarctica, and there was none. Day after day, United States and New Zealand military and civilian radio experts attempted to get through to any Antarctic station, but there was no response.

The silence in the south was as complete as though the whole region had suddenly evaporated. Even the United States Navy weather ship, stationed between the South Island of New Zealand and Cape Hallett, the northern promontory of Antarctica, could not reach the ice-cap stations.

The "storm" in the ionosphere, the electrified layer of the high atmosphere, was causing radio signals to be absorbed instead of reflected. Civilian and military scientists bound for Antarctica continued to pile up in Christchurch, a city of two hundred thousand New Zealanders who put out the welcome mat for Americans in World War II and have never withdrawn it.

At length single side-band communications were established with McMurdo Sound. They were intermittent, but enough weather data got through to give meteorologists a picture of conditions at McMurdo, and the Navy let the Constellation go.

En route it made radio contact with an Air Force Globemaster which had been relaying weather information from McMurdo to the Navy picket ship, which in turn was relaying it to Christchurch. The voice of the C-124 radioman came through in the cockpit of the "Connie," punctuated by coughs and crackles. He was reciting temperatures and pressures and times. In the midst of this droning recital, the Globemaster came into view, lumbering northward at lower altitude. When we arrived at McMurdo, after a flight of ten hours, the weather was clear and sunny.

The pole station continued to remain blacked out, however, for several more days. Not a peep came through. Then communications slowly opened enough to let a weather advisory slip through, and we were on our way to Amundsen-Scott aboard one of the new C-130 Hercules aircraft. As big as but faster than the Globemaster, the Hercules was equipped with skis as well as wheels, and it landed us smoothly at the south pole on a runway scraped by a tractor which had been parachuted down in 1959.

At the pole station, the men had heard the blow-by-blow account of the presidential election of 1960, except for recent tabulations which could have changed the slim lead held by the late President John F. Kennedy. The blackout had descended at this point, and the pole people hadn't heard a word since then. Their first question as we climbed out of the aircraft into the brisk polar wind at 38 degrees below zero Fahrenheit was, "Who got elected President?"

High winds that whirl snow and ice crystals make rescue operations hazardous in an Antarctic storm.

A ring of cyclones

Preliminary circulation studies[4] show that the interior of West Antarctica is stormier than the East Antarctic interior. A major storm path was found through Marie Byrd Land from the Ross to the Weddell Seas.

Cyclones moving along the coast of East Antarctica usually do not penetrate deep into the interior until they reach Wilkes Land. They are blocked by the presence of a semi-permanent anticyclone system over East Antarctica, until they have passed the 90th meridian, beyond which they move inland.

"The fact that the interior of eastern Antarctica is comparatively free of storms and cloudiness is one of the contributing factors in making this area one of the coldest regions on earth," Moreland reported.[5] "Since the elevation of this region averages 10,000 feet above sea level, clear skies permit a large amount of heat to escape from the surface through the dry, transparent atmosphere."

Moreland also noted that one of Weather Central's greatest difficulties was radio fadeouts. One major fadeout occurred each month and lasted three to five days. During this period radio contact with the outside world ceased, and sometimes even communication with other stations on the ice was blacked out. As we shall see in the next chapter, this phenomenon became a subject of a major study in itself.

In the eighteen-month period after July 1, 1957, when Weather Central went into full operation, regular observations were made four times a day at fifty stations on the ice cap, in the Panhandle, and on Antarctic and sub-Antarctic islands, totaling more than a hundred thousand observations. This mass of information provides a picture of a continental-oceanic exchange of air, which is continuous during the year but more intense in the winter, related to the low-pressure systems that race around the continent about three hundred miles out to sea. The circulation set up by these systems forces warm, moist air to rise over the edges of the continent and to penetrate into the interior. This air is the cause of most of the snowfall. But as it rises ever higher over the interior plateaus it becomes cold. Ultimately it merges with the stagnant high-pressure zone over the east plateau to form the "zone of permanent high pressure," which Murray predicted.

Becoming colder, the air settles and moves down the slope of the ice cap toward the coasts. When the flow reaches the sea, it is again caught up in the belt of cyclones whirling around the continent.

The storms divide when they reach the Ross Sea. Some continue on over the Pacific Ocean, others enter Antarctica and sweep across Marie Byrd Land and the Ellsworth Highland to the Atlantic Ocean.

With their network of weather stations in East Antarctica, the Russians have developed data showing that the winds have played an important part in shaping the ice cap. As the winds move up the East Antarctic slope from the sea, their moisture is squeezed out of them by chilling. Snowfall, therefore, is concentrated near the coast, for by the time the air moves far inland it has become quite dry.

One would expect to find less snow accumulation in the interior, but that is not so. All the seismic work by American, Russian, and French investigators shows that the cap is shallower near the coast, where more snow appears to fall, than inland, where there is less snowfall. In modern times, at least, the region of thickest ice does not coincide with the area of the heaviest snowfall. That is only one of the puzzles that meteorologists and glaciologists have yet to resolve.

The center of the great ice dome of East Antarctica is not on the polar plateau but on the eastern flare of the bell. At Vostok, midway between the coast of Wilkes Land and the pole, the ice is 12,136 feet thick, with rock bottom 656 feet below sea level, while at the geographic pole ice thickness is 9216 feet.

The "pole" of Antarctic wind circulation is not at the geographic pole either. It is farther to the north, toward Wilkes Land. From a scientific viewpoint, Antarctica has five poles. There is the geographic pole, where the meridians converge at the southern end of the planet's axis of rotation. Then there is the magnetic south pole, toward which the compass needle points, and where the magnetic meridians converge. There is the geomagnetic south pole, too, where the magnetic pole would be found if the earth's magnetic field was uniform—which it is not. Finally, there is the "pole of relative inaccessibility"—the point farthest inland from the coasts, at the mathematical center of the continent—the hardest to reach, and hence the most inaccessible place in all the Antarctic.

Not one ice cap but two

While the ice cap appears to be in balance now, the ample evidence that it has fluctuated locally at McMurdo Sound, in the Horlick and Pensacola Mountains, and on the Panhandle may mean there has been recession over the entire great mass. Unquestionably the ice may have diminished in some places, but it may also have increased in others, so that the total volume may be the same.

A theory which supports this conception of ice-mass transfer in Antarctica was proposed in 1959 by Wexler, then chief of research at the United States Weather Bureau and also chief scientist of the American Antarctic program during the IGY. Dr. Wexler proposed that there may actually be two ice ages in the Antarctic instead of one, and that their trends run in opposite directions. One glacial configuration, according to this view, is represented by East Antarctica, and the other by West Antarctica.

Wexler cited meteorological evidence[6] that snowfall on the continental cap of East Antarctica may occur at the expense of the West Antarctic ice sheet, and vice versa. He suggested that the East Antarctic cap reached its maximum about 10,000 years ago and has been diminishing ever since. During this time the ice cap in West Antarctica has been growing.

While ice recession on the Panhandle, which may be considered a part of West Antarctica, does not support this idea, the theory is based on IGY weather observations, which show that the present storm path courses through West Antarctica. Though they may not have done so in the past, the great snowstorms follow this track today. Consequently, West Antarctica now receives twice as much snow as the eastern continent does.

In prehistoric times, Wexler suggested, the storm track moved over the eastern continent. It built up the East Antarctic cap to elevations of 8000 to 14,000 feet, or 1.5 to 2.5 miles of ice. These elevations are found today on some portions of the eastern ice plateaus.

Now the meteorological effect of a pile of ice nearly three miles high was to deflect the storms into West Antarctica, which is lower in elevation. True, the ice is not as thick there, but it is lower for another reason. The sub-ice structure is sea bottom, a mile below the standard sea level. Most of East Antarctica is continental land.

According to Wexler's hypothesis, then, for the last ten thousand years, West Antarctica ice has been building up and East Antarctica ice has been diminishing because of the shift in the storm path. On this basis, ice recession in the McMurdo Sound region, at the border of East Antarctica, may be accounted for by the shrinkage of the East Antarctic sheet.

The theory of alternating ice ages in Antarctica does not require any change in solar activity. Wexler imagined a purely mechanical process: the East builds up until its very height acts to divert its source of nourishment to the West.

Dehydrated, so to speak, the east cap shrinks down for lack of precipitation, until its elevation no longer deflects the snowstorms to the west. The storm track then reverts back to the eastern plateau, and the western ice cap begins to dwindle.

Wexler believed this seesawing of ice ages fitted into a world climatic context. He told a subcommittee of the United States House of Representatives Appropriations Committee:

As you know, we had, perhaps, 30,000 years ago a rather warm interglacial period in North America, followed by the ice age that locally has been called the Wisconsin Age.

It reached its maximum some time around 18,000 to 20,000 years ago. Most of the ice has melted since that time, during which the sea level was raised by 200 feet.

These years I give you are quite comparable to the times that have occurred here in Antarctica. About 10,000 years ago, the ice started building up in West Antarctica at about the time, we think, that the ice reached its maximum in East Antarctica. There seems to be a very definite connection between events here governing ice ages in Antarctica and ice ages in the Northern Hemisphere.

The Weather Bureau research chief advanced his alternating-ice-age theory as a probability. He wanted to demonstrate how an ice cap could respond to meteorological trends. If one accepts the idea, it is a simple matter to attribute observable deglaciation to it, rather than to a general warming up of the region.

Wexler advanced several theories of this kind. To him, a theory was something of an intellectual game. "A theory isn't anything sacred," he said. "It's just a means of pulling observations together so that you can have something to argue about."

In this vein he proposed another idea, which he did not feel was incompatible with the first: in addition to having alternating ice ages, Antarctica is becoming warmer. He based this on a graph of nearly fifty years of temperature records at Little America, on the east coast of the Ross Sea—one of the oldest camp sites in the Antarctic. As I have related in earlier chapters, Roald Amundsen established his base Framheim there in 1911, for his dash to the pole. After he sailed home to Norway in 1912, the area was occupied by a series of Little America stations until 1960. Admiral Byrd developed Little America I on his 1929–1930 expedition, Little America II on his 1934–1935 expedition, and Little America III in 1939–1941. Then, from 1954 to 1959, the site was used as an IGY station and abandoned only after the Navy established its main supply base on Ross Island in McMurdo Sound, four hundred miles to the west.

Wexler plotted the annual temperatures which had been recorded at Framheim-Little America intermittently since 1911. Between that time and 1959 he found the average annual temperature had risen 4.7 degrees Fahrenheit—a substantial increase. However, the trend was based on only six years of observations spread over a forty-eight-year period. Some experts have doubted that the increase was produced by natural causes and argue that the heat emitted by the camp itself may have influenced readings. Wexler found this objection amusing.

Little America lies on 78 degrees south latitude. It is the same distance from the south pole as the island of Spitsbergen, at 78 degrees north latitude, is from the north pole. Two places equidistant from opposite poles on the earth's surface are called conjugate points, and such points play a key role in a number of Antarctic investigations.

Imbued with the idea of synchronization in hemispheric changes of climate, Wexler examined the Spitsbergen records. What had been going on there while Little America was getting warmer? He found that between 1912 and 1959 the annual average temperature at Spitsbergen had increased 11.3 degrees Fahrenheit, nearly three times as much as at Little America. He reported:

The Spitzbergen warming probably represents an extreme condition of a climatic change which is believed by some to encompass the entire Arctic basin. Whether Little America is a representative or an extreme example of a similar warming trend in the Antarctic cannot be determined in the absence of similar long and reliable records elsewhere on the continent.[7]

The Navy's decision to abandon Little America in 1959 ended the likelihood of continuing the measurements there.

The last clean milk

Once the gateway to Antarctica, Little America had been declining with the growth of McMurdo. Though a number of scientists considered it the best science site in West Antarctica, the Navy acted under other considerations. After five years the station was in such a state of disrepair that it required reconstruction if it was to be used much longer. Another factor was the instability of the ice at Kainan Bay, where the station was located. The famous Bay of Whales where Framheim and Byrd's Little Americas I, II, and III were sited had disappeared during the 1940s.

Early in 1963 portions of Little America III were found drifting in an iceberg in the Ross Sea, three hundred miles from the great ice shelf. A huge section of the shelf where the station had been built in 1939 had broken off, carrying sections of the station's structures with it. The discovery was made by a lookout aboard the icebreaker USS *Edisto*. He saw manmade objects protruding from one side of the great iceberg, which was four city blocks long and towered sixty feet above the sea surface.

Two men made a reconnaissance of the iceberg in a helicopter. They found walls of buildings and antenna masts of Byrd's old station, buried under twenty-five feet of snow. A freak accident had caused the iceberg to shear off the Ross Shelf at exactly the point where the station had been built.

Early in 1960 Little America V had contributed its final scientific offering to the world. A party headed by Captain E. E. (Doc) Hedblom, a Navy expert on polar medicine, went there to dig up cans of condensed milk which had been left by Byrd's 1939–1941 Antarctic Service Expedition.

After twenty years the milk was still well preserved in the south polar refrigerator. It was the only milk known to exist in the world that had been produced before the atomic age. Would it contain any strontium-90, cesium-137, or any other long-lived radioisotopes which have rained down upon the earth since the explosion of nuclear bombs began? From the time the first atomic bomb was exploded, the biosphere of earth, where man resides, has become contaminated by the fallout of radioactive debris, which has settled onto fields where it has been consumed by cattle eating the grass. Hence it turns up in their milk and beef.

Here in the ice at Little America was milk that might be entirely free of radioactive fallout. And indeed it was. It was the only "clean" milk left on the entire planet.[8]

The heat sink

Wexler emphasized, as Scott had many years before, that if the climate did warm up in Antarctica, the first probable result would be an expansion of the ice cap because of greater snowfall. The total volume of the ice would be reduced only after the climate had warmed sufficiently so that the loss of ice through wastage, ablation, and melting exceeded the snowfall.

Since a glacier is the product of combined cold and moisture, the simple lowering of the temperature does not produce one. And as G. Frederick Wright said, half a century ago, "Before an area can maintain a glacier, it must first get the clouds to drop down a sufficient amount of snow on it."[9]

If a planet-wide warming occurred, the ice caps on Greenland and Antarctica would at first become bigger. As they did so, the increased area of the ice cover would reflect more of the sun's energy back into space and thus tend to reverse the warming-up process.

The atmosphere is not warmed directly by solar radiation. About 72 per cent of the sun's energy reaching the earth in the middle latitudes is absorbed by the land and seas.[10] It is then reradiated as infrared radiation, or heat. The lower atmosphere is warmed in this way.

During the austral summer, when the sun shines twenty-four hours

copter view of the iceberg containing portions of Little America III.
wooden poles protruding from the top surface probably supported
station's antenna array. Embedded in the side of the berg, at center,
e wall of a building, approximately 16 feet wide by 8 feet in height.

a day, the south polar plateau receives more solar radiation than is possible anywhere else in the world, not only because of its geographical location but because of its elevation. Yet only 13 per cent is absorbed by the snow surface. Approximately 74 per cent is reflected back into space by the mirror-like ice sheet, and another 13 per cent is taken up by water vapor and dust in the atmosphere.[11]

Antarctica is a radiative heat sink. That means it dissipates more heat than it receives from the sun. In one year, according to Wexler's estimates, its heat loss to space equals the total United States electrical-energy production for five years. The only way it can achieve a temperature balance is by exchange of air with the warmer latitudes. If this could somehow be stopped, Antarctica would continue to cool indefinitely.

The Arctic too is a heat sink, but not nearly so large a one as the Antarctic. The ice caps of Greenland and Antarctica act as great mirrors, reflecting sunlight back into space. Quantities of heat transported poleward by the atmosphere and oceans are dissipated by the ice sheet too. It is not hard to see, therefore, why any major change in the ice sheets of the earth would affect the planet's surface heat balance. One might imagine that without the ice caps the earth would be warmer and swampier, as in Jurassic times, 125,000,000 years ago,

A disabled helicopter awaits help on top of Mount Discovery, a 9000-foot peak overlooking the shining white surface of ice called a heat sink.

with giant reptiles munching the tops of feathery trees in a landscape something like the bayou country of Louisiana. Sea level, of course, would be much higher than at present, and epicontinental seas would once more flood large portions of the American Middle West, the Southwest, and other low-lying areas on the continents.

A warmer earth

The indications of climatic change in Antarctica point to a long-range warming trend. There is also a consensus among climatologists that in the northern hemisphere the entire Arctic basin has been warming up in this century.

The thickness of the ice forming annually in the Arctic Ocean has diminished from an average of 144.69 inches at the time of Nansen's expedition in the *Fram* (1893–1896) to 85.83 inches during the drift of the Russian icebreaker *Sedov* in 1937–1940. The extent of drift ice in the Soviet sector of the Arctic has lessened by 600,000 square miles between 1924 and 1944. In West Spitsbergen the shipping season has been extended from 3 months in 1900 to 7 months in 1940.[12]

Cod have been migrating northward in larger and larger numbers. They are being caught for the first time, in modern times, by Greenland Eskimos.

In Norway, Sweden, and Finland, the timber line has advanced northward. The permafrost line has been receding in northern Canada.

Seven new southern species of birds have begun breeding in Iceland within the last sixty years and thirty-seven new species or sub-species have been added to the Icelandic list since 1938.

"Everyone is aware that our climate is changing," said John Clark, associate curator of sedimentary petrology at the Chicago Natural History Museum.[13] "Most people are aware that such temperate areas as Illinois are growing notably warmer." Clark points out that Chicagoans see cardinals and mockingbirds, which forty years ago never came north of southern Illinois. Opossums have moved as far north as central New York State. The maple-sugar industry of northern Ohio is dying out, he says, because the sugar maples cannot endure the warmer climate.

Clark and Erling Dorf, a paleontologist at the University of Michigan, have charted recent climatic variations in the northern hemisphere in the following way.

In the millennia 5000 to 2000 B.C. there were generally higher temperatures than there are today. This period was followed by a general cooling, which reached a minimum temperature about 500 B.C. There was then a subsequent rise in temperature.

A major warm episode occurred about 1000 to 1300 A.D. During this time, a colony of three thousand Norsemen, established by Eric the Red, grew crops and raised cattle and sheep in southwestern Greenland. Vineyards were growing in southern England.

The climate then changed. The Norse colony faced adverse weather conditions and without aid from Iceland or Norway it slowly died out. About 1600 the climate in Europe became cooler. Glaciers readvanced in the Alps, overrunning valley settlements hundreds of years old. This cooling led to the "Little Ice Age" from 1650 to 1850.

Since 1850, according to Clark and Dorf, the general climatic trend has been toward warmer conditions. Clark expects this to continue, with brief but temporary reversals, for another two hundred years at least—possibly longer.

Other observers have noted similar changes in our climate. The Swedish scientist H. W. Ahlmann, who was an ambassador to Norway, said that the glaciers remaining in Iceland were larger at the end of the nineteenth century than when the land was colonized between 870 and 1200 A.D.[14] For hundreds of years the glaciers dwindled. Then, in the seventeenth century, they began to grow. For three hundred years they were more extensive than at any time since the melting of the last remnants of the Pleistocene inland ice in the mountains of Scandinavia. Within the last seventy to eighty years, according to Ahlmann, these glaciers have been receding. They are now as small as or smaller than they were in Roman times, when the climate of western Europe was milder than it is today.

It is evident that the Norse colony in Greenland was the victim of a climatic change which affected that region before it spread to Europe. Excavations by the Danish government in the cemeteries at Jerjolfnes yielded heavy clothing from the year 1400 and bones warped by malnutrition.

The British scientist R. G. West proposed that after the ice left the mountains of Britain ten thousand years ago there was a warm period until about 3000 B.C. Ever since, he said, the climate has been deteriorating, with a drop in annual mean temperature of 3.6 degrees Fahrenheit.[15] Hence it is possible that we may be two-thirds of the way through an interglacial period. West suggested, however, that man today is living in a relatively stable time, in which nothing so dramatic as another ice age or the melting of the polar caps is likely to happen.

Climate in an ice age

Since we have barely a century of weather observations in the United States and less than fifty years of synoptic weather data for most of the planet, there is hardly sufficient meteorological information on which to base a climate trend. Simultaneous observations over the planet, including the polar regions, have been made only since 1957. We have just begun to make a global analysis of weather, and it is in this context that Antarctic meteorology becomes significant.

An ice age presents its own kind of climate. Like other climates, it is varied. For example, there is as much difference in winter temperatures between the high plateaus of East Antarctica and the "Banana Coast" of Wilkes Land on the Indian Ocean as there is between Minneapolis and Miami.

At 30 degrees below zero Fahrenheit, the Indian Ocean coastal region may still be 70 degrees warmer than the high ice fields of the interior, where 100 degrees below zero is not unusual.

Before the IGY, the full range of temperature in Antarctica was not known and was considerably underestimated. The lowest temperatures recorded on the ice between 1900 and 1955 were exceeded by readings of −90 degrees F. in northeastern Siberia in February 1933; −81.4 degrees F. at Snag Airport, Yukon Territory, Canada, in February 1947; and −86.8 degrees F. at a French station on the Greenland ice cap in February 1950. During the IGY, these records were regularly broken by lows at Amundsen-Scott Station at the south pole and by Soviet stations in East Antarctica.

During the eighteen months of the IGY, temperatures at Amundsen-Scott Station dropped below −100 degrees F. on five days. The lowest was 102 degrees below zero.

Even colder than the pole is the Russian scientific station Vostok. It is 1000 miles from the pole on the eastern ice in the Australian quadrant. Its elevation is brutal—11,200 feet above sea level, the equivalent of an altitude of nearly 13,000 feet in a temperate latitude so far as oxygen pressure is concerned.

On August 24, 1960, the temperature at Vostok reached an all-time low—126.9 degrees below zero Fahrenheit. It was the lowest natural temperature ever measured on the earth's surface up to that time.

Remarkable contrasts in temperature occur. On May 7, 1957, when it was −100 degrees F. at the pole, it was 30 degrees F., or just 2 degrees below freezing, at Little America, only 700 miles north. The cause of the 130-degree difference was an occlusion, a double frontal system in which a cold front overtakes a warm one. The temperature contrast across the two fronts weakened as they approached the pole. On May 11 the warm front, modified by its passage over the ice, was merely a remnant as it crossed the pole. But it still was intense enough to cause cloudiness and snowfall and to raise the temperature 22 degrees in one day.

As spring ends in the Antarctic, the stratosphere at the 11.4-mile-altitude level cools for eight continuous months. The temperature ranges from −25.6 degrees F. in mid-November to −127.3 degrees F. early in August.

When the sun returns in the spring, the stratosphere begins a four-month warming period. Investigators have been surprised at the rapidity with which it takes place: the south polar stratosphere changes from winter to summer temperature structure in thirty days—between mid-October and mid-November.[16] This is a warm-up of 140 degrees Fahrenheit, from the low in mid-August to a high in November. Cooling then begins at the 11.4-mile level in mid-November and continues through the summer solstice and into the long winter night.

Daily measurements in October showed that warming of the stratosphere was not steady but intermittent. There were a number of quick temperature changes. An explosive warming was observed in the last week of October, when a 68-degree rise in temperature was measured at a 16.7-mile altitude within 12 hours. With the warming, stratosphere circulation changes. After the explosive warming, a jet stream was observed traveling at 130.8 miles an hour.

The mechanism which accounts for this sudden warming up of the high atmosphere and its effect on the winds below has not been explained. We examine several theories about them in the next chapter.

Arctic versus Antarctic

A definition of a polar climate that seems to be acceptable to most meteorologists is one which sustains perpetual snow and ice, or in which the warmest month's temperature averages less than 50 degrees Fahrenheit, too cold for trees to grow.

On this basis, the south polar climate of the Antarctic covers three times the area of the north polar climate of the Arctic. According to Rubin,[17] the principal explanation for this is that the ocean surrounding Antarctica transfers incoming solar heat to deeper waters. Thus, the sea maintains a fairly uniform surface temperature, which minimizes the heating of the atmosphere by the ocean in summer. On the other hand, when the sunlight strikes the land around the perimeter of the Arctic, the heat is taken up within a few feet of the surface. This raises the temperature of the land and of the air above it.

The temperature difference was noted in 1911 by George Simpson, the meteorologist on the *Terra Nova* expedition. He compared average temperatures at Scott's Antarctic base at Cape Evans with those of a conjugate point in the Arctic at 77 degrees, 51 minutes north latitude. For 12 months of 1910–1911 the Cape Evans temperature averaged 0.4 degree F., while that of its Arctic conjugate was estimated to be 2.5 degrees F. But, he concluded, "The real severity of Antarctica is not shown in its low minimum temperatures but in its low maximum," since the north pole in July averaged 30 degrees F., while the December–January (summer) average at Cape Evans was 9 degrees colder at 21 degrees F.[18]

Like the Arctic, Antarctica is the region of the *kernlose* (core-less) winter—the winter that has no "bottom." The winter solstice, June 21, does not necessarily usher in the coldest weather. In fact, June is frequently a warmer month than May, July, or August. Most record Antarctic lows have been reported in August. Station records reveal that the temperature warms slightly in June and then becomes colder in July, so that there is no particular time when the meteorologist can say, "This is the bottom of the winter. From now on the temperature will rise."

One explanation is that the ocean around the ice cap does not freeze to its maximum extent until after winter has started. Consequently, warm air moving in from the north crosses a wider area of sea ice in late winter and hence is cooled more than in early winter.

The circulation of Arctic and Antarctic atmosphere streams are similar, but there are important differences. The effect of these regions on the weather of their respective hemispheres differs too.

Inhabitants of the northern hemisphere are familiar with breakouts of north polar air, which bring cold fronts racing down upon them in the winter. But such breakouts are relatively infrequent in the Antarctic, even though it is colder. Occasionally, however, Antarctic air does move north as far as southern Brazil.

In the Antarctic a ring of strong westerlies prevents any rapid exchange of air with the rest of the southern hemisphere. When the sun rises, however, the westerlies weaken, allowing warmer air from the north to flow into the glacial region.

During the winter in the Arctic, on the other hand, the cyclonic vortex, which tends to confine the north polar air within the Arctic Circle, breaks down frequently. This enables cold air to flow over North America, Europe, and Asia.

There is more to the story, investigators have found. While air exchanges in the winter between Antarctica and the northern latitudes is weak at the lower levels of the atmosphere, there is a good deal of exchange in the stratosphere. Evidence for this has been found in the presence of ozone in the high atmosphere over the Antarctic during the winter darkness. Ozone, a molecular form of oxygen containing three atoms, instead of two, as in the normal oxygen molecule, is a constituent of the atmosphere produced by ultraviolet radiation in sunlight. Consequently, if there is ozone in the stratosphere over Antarctica when the sun is not shining, it must have come from sunlit regions to the north. Ozone for years has been recognized as a good atmosphere tracer, especially of currents in the stratosphere. Even though it is manufactured in the high atmosphere by sunlight, its maximum concentration is found not over the equator, as one would expect, but over sub-polar latitudes.[19]

Ozone measurements were made at Little America and Pole stations during the IGY, with balloon-lifted detection devices. In 1962 rockets were introduced into the program by the Schellenger Research Laboratories of Texas Western College. The 8-foot rockets carried an instrument payload 40 miles high. Separated from the rocket, the instruments descended on parachutes and landed on the Ross Ice Shelf or the bay ice of McMurdo Sound.

During the IGY worldwide measurements showed that ozone was more abundant in the Arctic than in the Antarctic during the polar winters. Again, this suggests a greater air exchange with the lower latitudes in the Arctic than in the Antarctic during their respective winters. The greater exchange of air between polar and tropical regions in the northern hemisphere may be another reason why the Arctic is neither so cold nor so extensive as the Antarctic.

While Simpson and other meteorologists were aware in 1912 that the Antarctic was a more extensive and colder region than the Arctic, they had no means of determining the actual difference. IGY measurements showed the extent of the contrast. At the south pole monthly average temperatures range from −13 degrees F. to −80 degrees F. while at the north pole the range is 32 degrees F. (in July) to −31 degrees F.

The Arctic grounded ice, including the Greenland ice cap, is barely one-seventh of the Antarctic sheet. Hence, it reflects less solar radiation back into space and permits more to heat the ground. In summer, the Arctic Ocean pack becomes streaked with open leads of water.

A program of measuring the radiation reflectivity (albedo) of a large mass of ice was carried out in Antarctica during the IGY by two United States Weather Bureau meteorologists, Edwin C. Flowers and Kirby J. Hanson, at the south pole station. Flowers was chief meteorologist at Amundsen-Scott in 1956–1957 and Hanson succeeded him in this post in 1957–1958. Then, in 1959–1960, Flowers returned to Amundsen-Scott for another season, saying he had forgotten his slippers and had come back to look for them. He never found them.

The scientists used two sunshine gauges, technically known as pyrheliometers. One was mounted about 5 feet above the snow surface, 300 yards or so from the station. It automatically tracked the sun continuously during the 6-month polar day. The other pyrheliometer was mounted 15 feet above the snow surface and inverted in order to measure the snow reflection. The results of the experiment showed that 57 per cent of sunshine was reflected by the snow and 17 per cent by the atmosphere—a total of 74 per cent mirrored back to space. Of the remaining 26 per cent, half was absorbed by the atmosphere and half by the snow.

As a result of the snow reflection, the heat loss on the south polar plateau is large enough to cool the first 31 feet of snow by 100 degrees a year—if radiation alone controlled the heat balance. As we have seen, it does not. One proof of this is that the snow temperature varies less than 1 degree at a depth of 31 feet. The upper snow level must be getting heat from somewhere to balance what it is losing. The source is the air moving in from the north.

At Hallett Station, 1000 miles to the north of the pole, radiation studies showed that the amount of sunlight received there in clear summer skies was the same as that measured in summer at a station in New Mexico. In the summertime you can get just as good a sunburn on the Ross Ice Shelf as in Atlantic City. Of course, people don't go to the Ross Shelf for this purpose.

Yet the idea of vacationing in the Antarctic is not so fantastic. A group of imaginative Australians tried to interest investors in an Antarctic ski resort, and for a while the Argentine government ran excursion steamers across the Drake Passage to sub-Antarctic islands and the Panhandle.

The last desert

When one contemplates the enormous ocean of ice piled up on Antarctica, it is hard to realize that, meteorologically, the whole region is a desert. In terms of water equivalent, annual precipitation (snow, of course) ranges from 15.75 inches at Maudheim on the Atlantic coast to 2.75 inches at the pole. The annual average for the entire ice cap is hardly more than 4 inches. That is the annual precipitation of a desert.

Much of the snow one sees in the Antarctic is blowing—being moved from another place, not falling from the clouds. The tendency of buildings on the ice cap itself to become drowned in snow, as at Byrd Station and the Pole stations, also creates the impression of tremendous snowfall. But again, this accumulation around buildings is drifted snow. A building on the featureless ice cap is simply an obstruction to the free passage of snow as it is blown around by the wind. Snow accumulates until the obstruction is buried. At a campsite drifts 7 feet high are not uncommon. But a mile or so away the total new snow accumulation may be no more than 10 to 15 inches.

Compared to that in East Antarctica, snowfall in the west is fairly heavy—but only compared to the east cap. Little America recorded 84.2 inches in the 10 months from February to December 1957. But the water equivalent is barely 9 inches—about the annual average of Nevada. Compare it to the winter average snowfall from November to March of 104.3 inches at Caribou, Maine, or 77.2 inches at Duluth.

The wettest region of Antarctica appears to be the "Banana Coast" of Wilkes Land, so called because it is warmer than most of the ice cap. In July 1960, whereas the snowfall amounted to only 2 inches at Byrd and 4.5 inches at Hallett, Wilkes had 16.6 inches. Eights Station, the post-IGY base in Ellsworth Land, has had 24 inches of snowfall in the month of July, but this is equivalent to little more than 1.5 inches of water.

There is no typical Antarctic weather, but Little America is representative of the temperatures and snowfall of West Antarctica. A profile of the climate there was drawn in 1956–1957 by Ben W. Harlin of the United States Weather Bureau. He was a meteorologist at Little America that year, and later became scientific leader at Amundsen-Scott.

During February 1957, a midsummer month, the weather was mild, he reported.[20] The average temperature was 10.3 degrees F., which is

Above: *The roof of a building at Byrd Station sagging under a four-year accumulation of snow that caused the collapse of some installations.*

Below: *Loose snow is being scraped from the walls of the main tunnel for the new Byrd Station, constructed under the snow surface during the 1961–1962 austral summer, at a point about six miles away from its original site.*

fairly comfortable in an arid climate, especially when the sun is shining. Harlin noted that it was "not too unpleasant" to work outside.

As autumn came on, the temperature fell rapidly. During April it dropped to an average of −27 degrees F. July was the coldest month, with an average of −32 degrees F. Warming did not begin until the following October.

The worst blizzards came in May, July, and August. The longest storm lasted six days, from May 8 to May 15. The strongest blizzards blew at the end of August, when winds were clocked up to 75 miles an hour. When the winter ended, the buildings were nearly buried in drifted snow.

"Blizzards," said Harlin, "are the most unpleasant weather phenomenon in the Antarctic." No Antarctican who has endured the weird sonics of high winds against buildings day after day for a week will disagree with that.

And so, since 1957, the investigation of Antarctica has begun to fill in the last great blank on the global weather map. We now know something about the climate of an ice age. We know how cold it gets, how dry it really is, what its storms are like, and how it causes heat to leak away into space from an entire hemisphere.

We have learned through investigation how the physiography of Antarctica determines its climate in relation to world climate. We know it is much colder than the Arctic, much larger in extent, but less of an influence on hemispheric weather—or so it appears.

What is not yet known is whether and under what circumstances this glacial climate is changing, and what its trends are. Many years of observation on the ice will be necessary to tell that story. For if Antarctica is warming up, as Wexler's study indicated, major changes in climate must be in progress on the planet.

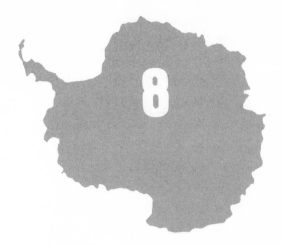

THE DAWN CHORUS

With the construction of Antarctic stations, the ice cap became a forward observation post for the investigation of the high atmosphere and of space. It provided the upper-atmosphere physicist with a place to stand in a polar region. And this was important for the study of phenomena occurring at the interface of the earth's atmosphere and the atmosphere of the sun.

The two atmospheres touch. One of the products of this contact is the aurora, with its fiery rays and draperies of green, red, and violet. Another is the ionosphere. Still another is the mysterious halo in the heavens called airglow.

Long before the IGY physicists had been speculating that the aurora was a visual manifestation of this contact, modulated by the earth's magnetic field. The field itself overspreads the earth like an invisible umbrella. It wards off all but the most energetic nuclear particles constantly bombarding the earth from the sun, except in polar regions. There the umbrella dips to the ground at the magnetic poles. Spiraling down magnetic lines of forces, charged nuclear particles can reach the atmosphere through the polar "holes" in the great shield of magnetic force. They can be counted and their effects measured.

The ice cap not only offered a place to stand under the "hole" but was an essential component for a global network of upper-atmosphere observation posts set up in the IGY.

199

The particle radiation of the sun consists of bits and pieces of hydrogen atoms. Some are positively charged hydrogen nuclei—solar protons. Others are negatively charged electrons which have been stripped away from the hydrogen nuclei. Similar particles come from other stars. Irrespective of where they come from, this atomic debris is called "cosmic rays."

The term is likely to be confusing. Cosmic rays are not "rays," in the electromagnetic sense, but particles. Nor are they entirely "cosmic" in the sense that they come from beyond the sun. Many do come from the sun, especially those in the lower-energy ranges. Cosmic rays are actually particles, bits of matter. They may be referred to as "corpuscular" radiation to distinguish them from light and radio waves and other electromagnetic energy, also emitted by the sun and other stars.

If the physicist thinks a particular array of particles comes from the sun, he refers to it as solar cosmic radiation. If he thinks the particles come from another star, he calls them galactic cosmic rays.

In any event, the best place on earth to observe this radiation is in a polar region. Antarctica is especially advantageous because the south pole is on grounded ice where the Amundsen-Scott Station stands. The north pole is not so handy. Since it lies in the Arctic Ocean, the upper-atmosphere physicist has had to float on sea ice to make observations from its vicinity.

Investigators of the high atmosphere set up their all-sky cameras, their cosmic-ray telescopes, neutron monitors, and ionosphere-sounding radios at stations all over the Antarctic continental ice cap. From the end of 1956 on, a steady flow of data has been accumulated.

It became apparent that the interaction of the solar and terrestrial atmospheres produced a complex array of effects. In addition to the aurora and airglow, there were disturbances of the ionosphere which caused the polar radio blackouts mentioned in earlier chapters.

On short-wave radios, investigators listened to strange whistles, hisses, tweaks, and bonks. These were sounds created by the interacting atmospheres and dispersion of terrestrial electrical energy along magnetic-force lines.

What was the significance of this investigation of the upper atmosphere? For the first time an international effort was being made to examine the frontier of the planet in space, and to sound the depths of space itself.

In the high atmosphere the great environmental investigation set in motion by the International Geophysical Year moved outward, from the terrestrial to the solar environment.

How high is high?

There is no marked physical boundary between the lower and upper atmosphere—one gradually becoming the other. Physicists think of the high atmosphere as the region above the altitude reached by radio-sonde balloons. The balloons ascend to about twenty miles above sea level and transmit data on temperature, pressure, and wind velocity to the ground by radio. A "radiosonde" is a sounding by radio. Above twenty miles, then, the atmosphere becomes the "high" atmosphere.

Although there may be no physical boundary, there is a physical difference between the lower and upper atmospheres. Because it is heated by the earth's surface, the lower atmosphere is in a constant state of turbulence. The transfer of thermal energy to the air is a driving force of weather.

For this reason, the gases which compose the lower atmosphere are well mixed. The relative concentrations are stable, with oxygen and nitrogen as dominant as at sea level. At high altitudes, beyond fifty miles or so, the mixing is weak. The heavier gases tend to settle out, and we find that the lighter gases, hydrogen and helium, are preponderant.

At about forty-five miles altitude, and above for four hundred miles, the effects of solar corpuscular and electromagnetic radiation produce the ionized or electrified region called the ionosphere. Here the gaseous molecules of air are shattered by the impact of incoming solar particles, which strip electrons away from nuclei to form positively charged ions in the thinning blanket of air.

It is the ionosphere which reflects longer radio waves back to earth. Otherwise, since the waves travel in a straight line, they would continue going out to space, as do the shorter radio waves used in television and radar. Global radio communications depend on the ionosphere. And when the regions of electrified gas are disrupted by events on the sun, varying degrees of communications blackout are imposed on the world, such as the one in 1960 (described earlier) which cut off radio communications with Antarctic stations for days.

The radio blackout of November 1960 was caused by what is called a "magnetic storm." The storm is a product of partial disruption of the earth's magnetic field in space by the sun's ejection of a stream of highly energetic electrified particles. Presumably the particles break through the disrupted field, more of them in polar than in other regions, and increase the ionization of the radio-wave reflecting regions. Instead of reflecting radio waves, the ionosphere then swallows them. Hence the communications blackout.

The magnetic storm of November 1960 was limited to polar and adjacent regions—the high latitudes. But more powerful storms can black out major portions of the world. One of these occurred February 10, 1958, the day after astronomers reported a big flare on the sun. About 28 hours later one of the most disruptive magnetic storms ever recorded struck the earth. At 9 p.m. on February 10, all direct radio communications across the Atlantic Ocean faded out. For a few hours radio technicians were able to route messages between Europe and America via South America and a big radio station in Tangiers. However, by 11 p.m. that day, radio communications between the continents were dead.

That night, aircraft flew ocean routes in silence. The voices, clicks, and beeps that fill the radio spectrum were gone. Fiery fingers of the aurora were seen in the southern United States, Mexico, and Cuba. Electrical currents raced through the high atmosphere and induced currents on the ground. In Canada and the northern United States power surges swept through transmission lines, tripped circuit breakers, and plunged communities into darkness, according to one report.[1] But, curiously enough, the effects were so diffused and the nature of the magnetic storm so poorly understood that the mass communications media for the most part remained unaware of what was going on.

After 9 p.m. that day, the transatlantic telephone cable was the only channel of communication open between Europe and America. But signals sent over it faded and swelled with surges of induced electrical currents with up to 2600 volts' potential.

For engineers and technicians who understood the situation, the experience was hair-raising. The blackout and the side effects continued until dawn, when the rising sun restored the ionosphere to its normal state.

In spite of the fact that the flare had been detected twenty-eight hours ahead of the storm, no official warning was issued. Scientists who might have warned the world were not in a position to do so. This gap has been closed. By 1964 the United States had set up a solar-flare warning system, but in the first months of its operation there was little occasion to use it. For 1964 was the Year of the Quiet Sun, when flares were at the minimum.

Infernos of the sky

Because the bombardment of particles from the sun on the atmosphere is heaviest at the poles, the high atmosphere above polar regions is much hotter than it is over the rest of the planet.

Rocket measurements of temperatures at altitudes of 100 to 125 miles over Fort Churchill, Manitoba, have reported temperatures of 4000 degrees Fahrenheit. At similar altitudes over White Sands, New Mexico, the temperatures were considerably cooler—1500 degrees.

It may seem strange to think of these altitudes, at which the Project Mercury astronauts orbited the earth, as being within the earth's atmosphere. However, in Project Mercury the capsules actually were in very thin, tenuous atmosphere. In fact, there was enough of it to slow a spacecraft down measurably and cause it ultimately to fall back to earth.

Recent satellite measurements indicate that the atmosphere extends as far out as 1000 miles, where it exists not as "sensible" atmosphere—air dense enough to be sensed—but in the fact that there is more gas in the region than in interplanetary space.

At 100 to 125 miles these blazing temperatures in the very high atmosphere would not be sensed as heat. The term refers to the energy of the gas molecules, which are excited by collision with incoming particles from the sun. Since more solar particles pour in through the polar holes in the magnetic field, this region over the polar expanses of the earth appears to have a higher density of energetic air molecules than any other above the planet.

Auroras are the visible evidence of the superheating of the very high atmosphere. They are the products of myriad collisions between the sun's particles and air molecules. Light is an energy product of this bombardment.

The geomagnetic field, believed to be created by a dynamo effect in the core of the rotating earth, sets up an umbrella of force from magnetic pole to magnetic pole. One can think of the incoming solar particles as a kind of cosmic rain falling upon this umbrella and rolling (or spiraling) down its slopes into north or south polar regions. This image is not precise, but it may serve to illustrate why auroras are usually confined in polar regions. The exceptions are when unusually energetic outbursts of particle streams, ejected by the turbulent sun, disrupt and crash through the magnetic umbrella to extend auroral effects to the lower latitudes.

From earliest times these electromagnetic displays, which have kindled awe and wonder in man for thousands of years, have been associated with the sun. Hence the term "aurora," from the Greek goddess of the dawn.

Investigations in Antarctica and in the satellite program have greatly clarified our understanding of the auroral mechanism. There is a tendency to think of it as the visual manifestation of the interplay of the atmosphere of the earth and the atmosphere of the sun.

One of the first observers to describe the aurora australis from within the auroral zone in Antarctica was Sir Charles Seymour Wright, when he was a young physicist on Scott's 1910–1913 (*Terra Nova*) expedition. For comparison he described the aurora borealis, which he had observed earlier in the Arctic:

Soon the low auroral glow becomes a quiet, greenish yellow arc, stretching across the sky nearly along a magnetic latitude. The first arc rises gradually toward the zenith, followed by a second, third, and perhaps even a fourth, moving in stately parade.

On a night of quiet aurora some or all of the arcs may pass the zenith about midnight and then go back, disappearing toward the axis [geographic] pole. During nights when the earth's magnetic field is disturbed, auroras become active just before midnight.

The arcs change into groups of discrete rays that arrange themselves parallel to the almost vertical magnetic lines of force. The general color remains a pale, greenish yellow, but often it brightens and shows a reddish tinge, particularly along the lower edges of the rays.

Even a vivid blue or apple green can appear intermittently.[2]

Sir Charles then wrote of the aurora australis, which he saw in 1911: "As in the north, a glow on the horizon was followed by the stately approach of quiet arcs. Since I was between the axis pole and the auroral zone, I saw the arcs moving inward toward the pole. Only rarely did they rise to the zenith and pass overhead. Afterward, they broke up into separate rays and draperies of an active aurora." In the fluctuations of these arcs and streamers, he said, one "senses the pulsations of a vast field of energy whose only source could be the sun."

A vast amount of auroral data collected since then has done nothing to change the accuracy of that visual impression.

The visible evidence of contact between the solar and terrestrial

Sir Charles Wright signing a visitors' log in the Shackleton hut at Cape Royds, on a return trip to Antarctica.

At left is the aurora study tower, used during IGY at Byrd Station.

atmospheres has never ceased to enthrall observers. E. J. Fremow, an auroral observer at the south pole in 1958–1959, described the aurora australis in this way:

The most spectacular displays were highly active draperies which nearly always moved to the zenith, where they were viewed end on as an inferno of violently moving red, green and blue. At times, the motion was so great as to blend the individual rays into a continuum of changing color, the red and blue combining to give a purple hue through which green rays could be seen rushing.

For the same year, there is this anonymous comment in the Amundsen-Scott Station reports:

I shall always be able to see in the eye of the mind the fantastic whirling of the beautiful, multi-colored auroral coronas, spinning overhead, as I shall remember the cleanness of the polar wind.

Thus do the auroras of the south continue to stir men of modern times.

The region of maximum Antarctic auroras was charted as north of McMurdo Sound and Little America, but over Ellsworth Station at one time and Wilkes at another. The northernmost reaches of large auroras were found between the South and North Islands of New Zealand. Most of the auroras during the IGY were seen in an ellipse about two thousand miles long, from the south pole to a point midway between Antarctica and New Zealand. However, the zone of greatest frequency shifts from the western side of Antarctica to the eastern side with the apparent motion of the magnetic field. The field seems to swing from one side of the ice cap to the other as the earth moves about the sun. Consequently the zone of maximum auroras shifts from Ellsworth toward Wilkes Station and back with the changing seasons.

Do-it-yourself auroras

Once the mechanism of the auroras was understood, it became possible to induce one artificially in the high atmosphere by exploding a rocket-lifted thermonuclear bomb (H-bomb) at high altitude. Theoretically, the release of energy would disrupt the magnetic field. It would allow charged particles to interact with molecules of air in equatorial and temperate regions, as they do at the poles.

Man's first "do-it-yourself" aurora was created August 1, 1958, when the United States exploded a thermonuclear device over Johnston Island in the Pacific Ocean. Minutes later a brilliant aurora was seen from Apia, Samoa, the first ever witnessed in the South Pacific.

Inhabitants of Hawaii were startled by the flash of the detonation. It was powerful enough to disrupt the ionosphere and the magnetic field. Both Mackay and RCA radio networks between California and Hawaii reported that communications were blacked out by the artificial magnetic storm for a day and one-half.

During the first week of August observers in Antarctica detected a new radiance in the sky at a wavelength of 6700 angstroms—the wavelength in the spectrum given off by lithium, never before seen in the high atmosphere. The first sighting was made at the American-New Zealand Hallett Station on Cape Adare. It was picked up later by all American Antarctic stations except Little America.

The presence of lithium in the upper atmosphere was attributed to the bomb. A second test in the vicinity of Johnston Island on August 12, 1958, produced less conspicuous effects.

Auroras were created over the Atlantic Ocean in another series of high-altitude nuclear-bomb explosions at the end of August and the beginning of September that year. The bombs were detonated at a 300-mile altitude in Project Argus.

The experiment had been proposed in 1957 by N. C. Christofilos of the Lawrence Radiation Laboratory, University of California, to test a theory that charged particles from the sun are trapped in the earth's magnetic field. The theory could be clarified, Christofilos suggested, by injecting charged particles artificially into the high atmosphere along magnetic-force lines. These particles should produce auroral effects by interacting with air molecules at conjugate points at either end of the magnetic-force lines. Observers saw the small auroras appear at each end of the magnetic-force lines in the North and South Atlantic Ocean, just as predicted.

The experiment revealed a part of the mechanism of auroras, whereby charged particles from the sun are funneled into polar regions. It enhanced a theory that solar particles, speeding toward the earth, become trapped in the geomagnetic field—a theory soon after to be confirmed by the discovery of the Van Allen radiation belts, clouds of charged particles caught in the magnetic field.

Once trapped, the particles spiral back and forth along magnetic lines of force, according to the emerging picture. In polar regions they spiral down the almost vertical force lines into the atmosphere. If, perchance, the particle strikes an air molecule, energy is released.

If not, the particle races downward until repelled by increasing magnetic-field strength. Then it spirals back to the opposite end of the great force field in the opposite hemisphere.

There, again, rising field strength will bat it back again, unless the particle strikes a molecule of gas. Back and forth the particles go, like shuttlecocks, until lost by collision.

The stage was now set for the discovery of the radiation belts later that year by Dr. James A. Van Allen and his research team from the State University of Iowa. The discovery properly belongs to the birth of the space age. It enlarged the picture of auroral mechanics, for the Van Allen belts were found to be the reservoir, the ammunition dump, of the particle bombardment. When the sun was active, the reservoir overflowed to produce heightened polar illumination.

With this model of how particles behave in the magnetic field, it became possible to produce a do-it-yourself radiation belt, in addition to an artificial aurora. Both were achieved July 9, 1962, with the Starfish Prime bomb test of Operation Fish Bowl. Again, the scene was the high atmosphere over Johnston Island. Again thousands of people in Honolulu watched the night sky light up green, yellow, and blood-red. And, again Polynesians in Samoa and Tonga saw the aurora.

Within a few hours the radiation-detector satellite Injun I reported the formation of the new radiation belt. The American Telephone and Telegraph Company's Telstar I, launched into orbit the day after the test, picked up very high intensities of the artificial radiation, much higher than had been anticipated. This was one of the less well-publicized functions of Telstar, the first communications satellite. Three other satellites, the Navy's Transit IV B, Traac, and England's Ariel, were damaged in varying degrees by the radiation.

Whistler's father

The image of the magnetic field elicited by these experiments and by nuclear-weapons tests was verified by a series of very-low-frequency (VLF) radio experiments, which formed a part of the upper-atmosphere research program in the Antarctic.

The VLF program attaches significance to what radio fans of the 1920s called "static." It is static only to the uninitiated. To the initiated, these strange and weird sounds are "atmospherics," or, in the jargon of the experimenter, "sferics."

Radio amateurs plow through these noises as they search the

spectrum for someone to talk to. A long time ago, however, it was realized that "sferics" are meaningful noises, emanating from the ionosphere and demonstrating the electromagnetic activity in that region. Sferics in the audio-frequency range, which you can hear, are principally hiss, dawn chorus, and other sounds such as a tweek, a clink, and a bonk.

The dawn chorus is heard as a series of short, distinct musical tones, in the frequency range of 1 to 5 kilocycles, rising or falling in pitch, and overlapping each other. Hiss, of course, is just hiss. Tweek, a sharp, musical chirp, is thought to be a dispersion of energy at 1800 cycles. Bonk and clink sounds are believed to be produced by the electromagnetic side-effects of thunderstorms.

Studies made at the south pole in 1960 suggest that hiss accompanies auroras. The "hissing auroras" were described by Henry M. Morozumi of the Stanford University Radio Science Laboratory, who correlated the intensities of the very-low-frequency radio waves which sound as hiss with those of auroral light during the austral winter from April 10 to September 1, 1960.

According to Morozumi's report, his observations showed "a close statistical association" between hiss and auroral intensities.[3] Morozumi was able to conduct the study without interference from sunlight for three months—the most conspicuous advantage of an auroral-observation post at the axis pole.

Henry M. Morozumi adjusting an oscilloscope at the South Pole Station.

Morozumi noted that the auroras visible from Amundsen-Scott Station reached peak intensity twice a day, at noon and midnight, and so did the VLF hiss. The correspondence was not quite the same, however. Whereas auroral light peaked during the break-up of the ray pattern, the principal form of auroras at the south pole, the loudest hiss came shortly before the rays broke up.

Morozumi described his hissing auroras in this way. In the beginning, a faint trace of hiss starts to appear on a recorder. It continues to be weak and intermittent for a while. At this time, the aurora appears as a glow or as faint bands to the magnetic north. Then the intensity of hiss rises and the aurora changes. Fine, swiftly moving raylike structures appear. The intensity of the hiss increases with the brightness of the aurora and the motion of the raylike structures. Extremely loud hiss comes with auroras having pinkish lower borders. Within five to fifteen minutes the forms break up into rays and draperies, advancing rapidly southward. Finally, as the rays break up and the light scatters, the hiss vanishes.

There is a strong possibility, Morozumi concluded, that the aurora and VLF hiss are caused by the same mechanism in space. He speculated that the ionosphere might interfere with exact observations of the two phenomena, for, although this ionized layer of the atmosphere is transparent to auroral light, it could cause some dispersion of the VLF waves of hiss. Morozumi recommended that a satellite be used to measure hiss above the ionosphere while observers on the ground watched the aurora and made simultaneous recordings of hiss at ground level.

By far the most fascinating and useful of the sferics are whistlers. Most of them sound like wolf whistles, starting out even and then descending in pitch. Some are heard also rising in pitch. These atmospheric sounds were first reported in telephone experiments in Austria in 1886. A scholarly report of a six-year investigation into whistlers, published in 1893 in Vienna, suggested that they represented electromagnetic phenomena of some kind in the atmosphere.

The "father" of whistlers, however, was Professor Heinrich Barkhausen, a German physicist and expert on magnetism, one of the chief scientists of the German High Command in World War I. Barkhausen's main mission in 1916 was to perfect methods of eavesdropping on Allied telephone conversations in the trenches across No Man's Land. He used a sensitive audio-wave amplifier which he connected to a pair of rods sunk into the ground. With it, he intended to pick up leakage of electrical current from Allied telephone wires and thus overhear the conversations.

Barkhausen's method worked well enough, and the resulting German intelligence might have confounded the Allies, had it not been for the frequency with which the military talk was drowned out by a shrill descending whistle, like the sound of an artillery shell in full flight. After a series of painstaking tests, Barkhausen concluded that the whistlers were not being produced by faulty equipment. Where they came from remained a mystery for many years.

It was not until the 1950s that Barkhausen's electronic eavesdropping methods and equipment were vindicated. Investigators in Great Britain and the United States found where whistlers came from. They were produced by long-wave electromagnetic energy generated by lightning strokes. Why couldn't Professor Barkhausen have figured that out? Well, for one reason, a lightning stroke that produces a whistler frequently originates in the opposite hemisphere.

The lightning origin of whistlers was confirmed in experiments by L. R. O. Storey at the University of Cambridge, England, and by M. G. Morgan at Dartmouth College, New Hampshire. In a nearby thunderstorm the lightning stroke is heard in a radio receiver as a click. Then, a fraction of a second later, the whistler is heard. The energy of the stroke has made a round trip to the opposite hemisphere and back again. The investigators deduced that the energy heard as a click traveled upward through the ionosphere and along a magnetic-force line into the opposite hemisphere. As it moved at the speed of light, the bundle of frequencies comprising the click was unraveled, so to speak, in the ionosphere.

Just as white light fans out into a spectrum when passing through a prism, so the click frequencies became spread out, like a troop of Boy Scouts at the end of a hike. The highest frequencies, traveling the fastest, were in the lead, and the lower ones tagged along behind. If the troop of frequencies became well separated, the listener would hear them as a whistle descending the scale.

It was reasonable to suspect, then, that the whistle was the echo of the click, but not all clicks were followed by whistlers. When they were, it was calculated, the click had traveled along a magnetic-force line into the opposite hemisphere and back again, returning as the elongated whistle.

When the velocities of frequencies in the click were timed, investigators found the energy from the lightning had made a round trip of 15,000 miles. The whistlers went far beyond the earth's atmosphere, and from this physicists were given a glimpse of the extent of the geomagnetic field in space.

Once the whistler rides the force lines, it bounces back and forth between the hemispheres, just as radiation particles do. One click may be followed by a train of whistlers, each succeeding one weaker than the one before as the energy is depleted by the trips through the ionosphere, until the whistler fades away.

When the characteristics of whistlers were known, it became possible to use them to sound the ionosphere and the magnetic field, in much the same way seismic shock waves of earthquakes are studied to sound the depths of the earth. One type of whistler made an excellent space probe—the double or "nose" whistler, which both rises and falls in pitch. On a graph, the energy-distribution curve of such a whistler as it passes through the ionosphere looks like the figure of a nose.

Early in the IGY, researchers at Stanford University found that the shape of nose whistlers indicated the intensity of the magnetic field and the flux of charged particles in the high atmosphere. They served as combination thermometers and Geiger counters. From analysis of nose whistlers, it was possible to figure electron density in the ionosphere and for a considerable distance above it—in fact, out to nearly 20,000 miles.

Whistlers thus became an important new probing tool. They enabled scientists to determine, for example, how activity observed on the sun affected the magnetic field and the ionosphere. All one needed to operate this auditory probe was a good short-wave radio.

Antarctic stations became key observation centers in a worldwide whistler program, as well as for other low-frequency-radio-wave phenomena. The stations were right where the magnetic-force lines, arching thousands of miles into space, came down toward the earth's surface.

Stanford University set up a "Whistlers West" (for western hemisphere) program to observe these and other atmospherics at conjugate points in the southern and northern hemispheres. Stations were located at Dunedin, New Zealand, and its conjugate, Unalaska, Alaska; the Macquarie Islands between New Zealand and Antarctica, and Kotzebue, Alaska. These stations kept a record of electron-density changes in the ionosphere by recording whistlers and other very-low-frequency emissions in the band between 400 and 30,000 cycles.

In 1960–1961 Stanford and the Pacific Naval Laboratory of Canada teamed up to conduct observations of ultra-low-frequency radio waves, which are below whistler frequencies, ranging down to a three-hundredth of a cycle per second—a cycle that lasts for five minutes. These long-period waves have been identified as "micropulsations" in the geomagnetic field. They follow the lines of force. Conjugate stations for this study were Byrd in West Antarctica and a Canadian listening post at the mouth of the Great Whale River at Hudson Bay.

Associated with the Canadian end of the project was Sir Charles Seymour Wright, who during the IGY and thereafter was working as a physicist with the Canadian Defence Research Board. Sir Charles said he suspected that micropulsations in the magnetic field and auroras were connected.[1] At the time the pulses were observed at Great Whale Station, the aurora borealis was active. At Byrd Station, the Antarctic conjugate point, no aurora could be detected. The sun was shining there.

These phenomena have become formalized in scientific literature as "magnetotelluric micropulsations." One can think of them as a "shimmy" in the geomagnetic field. It is as though the planet were breathing. What induces the oscillations in the field is a subject of wide speculation. Suggestions range from the mechanical effect of the earth's rotation to solar emissions.

A curious discovery about shimmy was made by the United States Air Force Office of Aerospace Research early in 1964. The agency in 1963 had set up a three-station network to record pulsations at Trinidad, Puerto Rico, and Austin, Texas. The Air Force scientists were concerned with long-period waves oscillating at 0.01 to 0.25 cycle a second. These were detected mainly during daylight hours, and seemed to involve the sun as the cause. But of even more significance was the fact that the waves were recorded simultaneously at Puerto Rico and Austin, about 2000 miles apart. The Air Force group has concluded that the waves represent pulsations of the entire magnetosphere—the full magnetic field of the planet.

Ions over the south pole

Several investigators have suggested that the ionosphere is involved in magnetic-field shimmy. It may represent a three-way interaction of solar radiation, the upper atmosphere, and the magnetic field.

These relationships do not represent new ideas by any means. The whole matter of the magnetic field, for example, has intrigued mariners since the advent of the compass.

It might be helpful at this point to review briefly the historical background of these concepts, some of which were discussed in Chapter 1. William Gilbert published his hypothesis that the earth was a global magnet in 1600. The voyage of James Clark Ross had for its scientific rationale the desire to find the south magnetic pole, which Gauss had predicted would be in latitude 66 south and longitude 146 east. As early as 1838 the German geographer Baron Alexander von Humboldt had proposed to the Duke of Sussex, president of the British Royal Society, that a chain of magnetic observatories be established around the earth.

Late in the nineteenth century the Scottish physicist Balfour Stewart attributed fluctuations in the magnetic field to an electricity-conducting layer of air in the upper atmosphere. He proposed that variations in the electrified layer induced variations in earth magnetism. But the significance of this idea was not realized until 1901, when Guglielmo Marconi sent wireless signals across the Atlantic Ocean. How could radio signals, propagated in straight lines, travel around the curvature of the earth?

Arthur E. Kennelly in the United States and Oliver Heaviside in England added to Stewart's concept of an electrified layer of air the idea that this layer reflected radio signals. Hence, the region now called the ionosphere became known as the Kennelly-Heaviside Layer.

In 1925 Edward V. Appleton and M. A. F. Barnett in England found that radio waves broadcast from a distant station returned to earth at an angle from somewhere in the heavens. Later that year Merle Tuve and Gregory Breit of the Department of Terrestrial Magnetism, Carnegie Institute of Washington, determined where the Kennelly-Heaviside Layer was. Working with the Naval Research Laboratory, they recorded and timed the echoes of radio waves transmitted 8 miles away. Their work showed that the ionized region of the atmosphere did indeed exist at an altitude of less than 100 miles.

With the development of worldwide radio communications, the ionosphere has been of intense commercial, as well as scientific, interest. The conception of its behavior and origin has evolved steadily, as a product of observation and advances in nuclear and astrophysical theory.

The ionosphere is the product of the atom-busting effects of solar radiation. The concentration of ions in the atmosphere increases to a height of about 150 miles during daylight. However, the entire electrified region extends outward for some 300 miles.

Because of the varying densities of ions at different altitudes, the ionosphere has been divided into several layers, or, more properly, regions. The lowest is the D region, at about 40 to 45 miles. Above is the E region, extending from 60 to 95 miles; then comes the F region, rising to 150 miles; and still higher is the F-2 region.

The density of electrons ranges from 120 per cubic millimeter of air in the E layer to about 450 in the F-2 layer. Now and then a strange temporary region appears that can reflect very-high-frequency radio waves, which normally pass right through the ionosphere. It is called "sporadic E" and when it appears FM radio and television signals can be detected hundreds of miles from the transmitters. More than once Antarctic radios have picked up the voice signals of European television stations.

These regions of ionization rise and fall in response to sunshine, to the tidal pull of the sun and the moon, and to changes in the magnetic field. The resulting cyclic movement produces strong electrical currents in the high atmosphere, which can affect the magnetic field at the ground level.

It is probable that the "dynamo currents" created by the seething of the region may be the origin of the electrical surges in telephone and power lines during magnetic storms. It is then that the restless sea of ions above us becomes mightily disturbed.

Ultraviolet light and solar X rays sometimes build up high concentrations of electrons in the D region. When this happens, radio signals are bent or absorbed once they reach the D layer, and are not reflected back to the ground. They may fade out entirely. The effect disappears at night, when the signal strength of radio transmission usually rises, compared with daytime. When the sun is shining there is always some ionization in the D region which reduces signal strength.

As part of a worldwide network of ionosphere-research stations, observation posts were established during the IGY at Amundsen-Scott, Little America, Byrd, Ellsworth, and Wilkes Stations.

During the austral winter of 1958 a rather surprising and certainly far-reaching discovery about electron density in the F and F-2 regions was made by observers at the south pole. They found that in July, in the dead of the southern winter night, electron density in the F regions above the pole was higher than it was anywhere else over Antarctica.

Physicists had expected that when the sun was below the horizon electron density would be low in the upper atmosphere. Yet midwinter soundings at the pole showed there was ionization down to an altitude of 160 miles. This was baffling, since the sun's rays do not strike the polar atmosphere below 280 miles in midwinter.

The persistence of the F region during the long polar night was new to ionospheric physics. But then, observations had been limited before the IGY to soundings at bases on the periphery of the continent. With regular observations at the pole, a more complex behavior of the ionosphere was beginning to appear. How could ionization take place in the high atmosphere in the absence of sunlight?

There were three possibilities. One was that the ionized regions might have drifted over the pole, as ozone does, from the lower, sunlit latitudes. A second possibility was that the solar particles bombarding the high atmosphere were being funneled into polar regions by the magnetic field. The third was that the ionized zones were being diffused along magnetic lines of force from the sunlit portions of the planet.

The theory that the ionized layers drifted over the pole was supported by observations showing daily variations in the F region, just as though the sun were shining on it. But there was a curious tie-in with the aurora too. Studies for one year, from June 1, 1957 to May 30, 1958, showed that ion density in the F layer during the winter darkness reached a peak when the auroras were most active. Now there was a definite correlation between auroras and the ionosphere.

But it remained for the development of space satellites to indicate what it might be. The discovery of the Van Allen radiation belts suggested that these regions of trapped particles might occasionally overflow with new infusions of nuclear debris from the sun. Such an overflow, it was theorized, might trigger the aurora and produce strong currents racing through the ionosphere.

One proponent of the theory, Joseph W. Chamberlain of the University of Chicago's Yerkes Observatory, calls the process "dynamic dumping." He has proposed a mechanism whereby superheated, ionized gas, called plasma, emanating from the sun, causes the Van Allen radiation belts to overflow. The belts dump their "excess" particles into polar regions with such high velocity that the aurora reds, greens, and violets appear as the energy product of the impact of these particles on those of the high atmosphere.

All the Antarctic stations have played a part in revolutionizing knowledge about the nature of these great electrical tides of the high atmosphere and of the interplay of solar and terrestrial gases. During the IGY several attempts were made to measure the changes in corpuscular and electromagnetic radiation from the sun by its effect on the ionosphere. Late in 1963 an international program of monitoring the ionosphere was set up at American, British, and Russian stations on the ice cap—the first experiment of this kind in history.

Sounding the ionosphere by radio waves is based on the physics of radio-wave reflection in the region. The effect is created by the free electrons. When a radio wave flashes into the ionosphere, it causes these electrons to swing back and forth. Each electron set in motion acts as an oscillator, generating a wave of the same frequency. Some of the new wave energy follows the original wave into space, and some is radiated earthward. This is the "reflected" signal. The original wave is not actually reflected, but is reradiated by free electrons.

Researchers before the IGY determined that the number of electrons that must be present to "reflect" a radio signal is directly proportional to its frequency. In other words, electron density must equal 12.4 times the square of the frequency as expressed in numbers of megacycles (million cycles per second).[5]

For example, a radio wave with a frequency of 3 megacycles would be reflected only when the electron density equaled $3 \times 3 \times 12.4$, or 112 electrons per cubic millimeter. In fact, 3 megacycles has appeared to be the upper limit of radio-wave reflection. That is why the very-high-and ultra-high-frequency waves of television go right through the ionosphere, except for freak reflections by the sporadic E region.

Because of the relationship between frequency and density of electrons, it becomes possible to measure electron density in the ionosphere by using radio waves of different frequencies. Antarctic stations are equipped with radio transmitters and receivers called ionosondes, which can measure both density and altitude of the electron layers.

One of the most ambitious programs of ionospheric soundings was developed before the IGY by New Zealand, whose territories and dependencies extend from the equator to the south pole. New Zealand set up a network of stations to make vertical soundings of the ionosphere by radio at Rarotonga, near the equator; at Christchurch, on the South Island of New Zealand; and at Campbell Island, between New Zealand and Antarctica.

In 1958 New Zealand built Scott Base, three miles from the United States Naval Air Facility on Ross Island. This extended the network to within seven hundred miles of the south pole. The entire Ross Sea region has been claimed by New Zealand as its Ross Sea Dependency, a pie-shaped wedge of Antarctica from the coast tapering to the pole. Ionosphere research was also carried on by New Zealand observers with Americans at the jointly operated Hallett Station at Cape Adare.

The New Zealand network made it possible to test a theory, held at the beginning of the IGY, that ionization in the high atmosphere was produced by monochromatic radiation from the sun. Observatons by the network, and by other national stations, did not bear out the idea.[6]

Scott Base researchers also made careful records of the VLF emission, or sferics. They found no particular correlation between the frequency and loudness of whistlers and other geophysical activity. In fact, the reverse was true. Whistlers were heard more frequently when the magnetic field and ionosphere were relatively undisturbed.

Forward scatter

In 1959 physicists attending an international Antarctic symposium at Buenos Aires, Argentina, suggested that simultaneous soundings of the ionosphere should be made at stations located at widely separated points on the ice as a means of keeping track of solar cosmic-ray activity. Since observers at most stations were making fairly continuous observations, the need for a coordinated program was not pressing. But advances in knowledge about solar-terrestrial relationships, especially in cosmic-ray theory, made some kind of joint observation effort on the ice cap a more interesting prospect.

For a number of years it was thought that all cosmic rays consisted of protons, but studies in 1963 found high counts of electrons. Early cosmic-ray theory also held that the origin of these particles was the stars, as the name suggests. Up to the middle 1950s, in fact, it was supposed that the production of cosmic rays by our own sun was a relatively rare event.

However, quite early in the Antarctic program neutron monitors, measuring the intensity of cosmic rays at the ice-cap surface, reported that cosmic rays fluctuated with changes in solar radiation. The monitors, which are radiation counters turned skyward, measure the number of times an incoming particle disturbs an electrical field. Cosmic-ray observations over a complete, eleven-year solar cycle, in both polar and temperate regions, indicated that cosmic radiation was emitted by the sun in energy ranges of millions of electron volts. This radiation could be detected more readily by balloon-borne instru-

National Bureau of Standards scientist ascends a 180-foot tower at Byrd Station that transmits signals by forward scatter to McMurdo Station.

ments, since most of the particles were too low in energy to reach the bottom of the atmosphere.

It became necessary, then, to differentiate between solar and galactic cosmic rays, and it was supposed that galactic radiation had the higher energy—up to 1000 billion electron volts.

During the declining phase of the solar cycle, between 1959 and 1962, sharp increases in the intensity of cosmic rays reaching ground stations were observed eight times in three years. For a quarter of a century before this three-year period, only five similar increases had been recorded.[7]

One sudden increase of cosmic rays reaching sea level occurred in 1961 on July 18 and another two days later. It was recorded at cosmic-ray monitoring stations operated by the Bartol Research Foundation, Swarthmore, Pennsylvania; at Thule, Greenland; and at McMurdo Sound. The hour-by-hour counts of cosmic-ray intensities at Thule and McMurdo between July 11 and 20 that year are almost identical. July was an active month on the sun, and flares were observed on July 11 and 17.

Further observations have only strengthened the evidence that disturbances on the sun are connected with increases in cosmic-ray intensities. However, a curious paradox exists here. There is also a marked decrease of cosmic rays associated with flares and sunspots. The decrease appears at the beginning of the event, and the increase follows it.

For a number of years it was suspected that the sun acted as a modulator or regulator of cosmic radiation, but just how it did so was unknown. The intensity of cosmic rays was observed to be inversely proportional to solar activity on one hand and directly proportional to it on the other. More recent findings by satellites suggest that the cosmic radiation is "swept out" of the inner solar system by emissions of hot ionized gas, or plasma, from the sun. Following the passage of the plasma cloud, the cosmic radiation reappears.

During the IGY and post-IGY period, cosmic-ray studies were carried on at the end of the active phase of the solar cycle. A question that loomed large to physicists was how the cosmic radiation pattern would change when the sun was quiet. Would the stream be steady? Or would there be smaller variations that had been obscured by the active part of the cycle? At the end of 1962 Great Britain, the Soviet Union, and the United States agreed to investigate the problem by using their resources in Antarctica.

Cosmic-ray particles are associated, along with ultraviolet and X rays, with electron production in all regions of the ionosphere. The joint proposal was to monitor continuously the flux of solar cosmic rays by observing their effects on the ionosphere. In turn, the effects were to be measured by changes in the ionosphere's ability to scatter radio waves. The focus of the experiment was on cosmic rays with energies in the millions of electron volts, or medium-low energy ranges. These particles penetrate to the D, or lowest, level of the ionosphere. It was therefore necessary to sound this level by observing changes in the scattering of the radio-wave energy.

This kind of investigation can be made only over large distances. It requires a means of transmitting the radio signal on a given frequency at one place and a means of receiving and analyzing the signal at another place, hundreds to thousands of miles away. The network of international science stations in Antarctica was admirably suited for such a project.

The three-nation cosmic-ray investigation got under way in December 1963. Aluminum girders, pre-cut to be assembled on the ice as antenna masts, were shipped to McMurdo Sound and to the British station at Halley Bay via London. The antennae were erected at three American stations—McMurdo, Byrd, and the south pole; at the Soviet Vostok station in East Antarctica, and at Halley Bay on the Atlantic side of the ice cap.

A signal beamed at the D layer from McMurdo would be propagated in the forward-scatter mode to Vostok, which would receive and analyze it. Vostok is about 1200 miles from McMurdo. Other paired stations are Byrd-Vostok, Byrd-Pole, McMurdo-Halley Bay, and Pole-Halley.

The height of the towers determines the pathways of the radio signals to the ionosphere and the obliqueness of their reflections from the D region. The National Bureau of Standards (NBS) worked out careful height specifications for the tower array at each station, including the British station at Halley Bay, and Vostok. But the NBS

At McMurdo Station the cosmic-ray laboratory is equipped with neutron monitors. The 30-foot tower at right transmits high-frequency radio waves by the forward scatter method from McMurdo to Vostok Station.

recommendation was not followed at Amundsen-Scott for a reason which had not occurred to project designers. At the south pole the Navy Seabee construction force has a rule that no man will work at an elevation of more than 180 feet above the ice surface. Consequently, the Seabees reduced the tower height at the pole from the NBS recommendation of 217 feet to 180 feet. Navy engineers insisted it was not safe for a man to work higher than that on the wind-swept ice, at 9200 feet in elevation.

NBS recommended a 217-foot tower for Halley Bay, two towers of 96 and 177 feet at Byrd Station, and 217-foot towers at McMurdo Sound and Vostok. A United States team was flown to Vostok to erect the aluminum towers there.

The heavenly glow

In contrast to the kaleidoscopic masses of light of the auroras is the soft, faint glow that is seen on a clear, moonless night, away from city lights. This is airglow, and it too is the visible evidence of the impingement of the solar atmosphere upon that of the earth.

Observations of airglow in both Arctic and Antarctic regions have identified at least four wavelengths of light, coming from the excitation of air molecules by incoming particles from the sun. There is a green line at 5577 angstroms, emitted by atoms of oxygen, and a red one at 6300 angstroms, also emitted by oxygen at a higher level of excitation. There is the yellow sodium line at 5893 angstroms, and the very powerful, but invisible, radiation in the infrared range of the spectrum, from a hydroxyl (OH) molecule.

Airglow is so faint that it is ordinarily invisible to the eye; yet it supplies a surprising amount of diffused illumination to the night sky. Observers of the Arctic Institute of North America have been making a long-term study of both aurora and airglow. They photograph and analyze the spectra of airglow in the night sky and during the long Antarctic twilight at the South Pole, Byrd, and Eights Stations—this last being a post-IGY station in the Ellsworth Highland northwest of Byrd. The significance of these observations is that a fuller understanding of what airglow represents provides us with a better understanding of the physics and chemistry of the high atmosphere.

While the mechanism of airglow has been considered for several years as similar to that producing the aurora, a different origin has been suggested recently. Studies at the Kitt Peak National Observatory at Tucson, Arizona, indicate that ultraviolet radiation from the sun, rather than the collision of solar particles with air molecules, may cause part of the airglow. Specifically, the ultraviolet causes the removal of an electron from oxygen molecules and atoms during the day. At night the free electrons recombine with the "parent" atoms, emitting radiation in the process.[8] Airglow is, again, a product of solar-terrestrial interaction, via electromagnetic or corpuscular radiation, or both.

At the outposts of science on the south polar ice, men listen and watch, photograph and record. In the austral winter night, the heavens are alive with light and sound from space.

ON THE FRINGE
OF THE
LIFELESS LATITUDES

The most remarkable line of biological demarcation on the surface
of the earth is the Antarctic convergence. It is a sinuous, oceanic
boundary, encircling the continent and nearby islands between 48
and 60 degrees south latitude. At the convergence the cold waters
of the Antarctic ocean plunge below the warmer waters of the
northern seas. This happens precipitously. Chilled by the Antarctic
ice, the southern ocean is denser and heavier than the tropic-warmed
seas to the north.

South of the convergence, or "divergence," as the boundary was
called by the late Harry Wexler, the lands are lifeless, except for
lichens, wingless insects blown in from the north, and the coastal
fauna which derives its sustenance from the sea.

The convergence represents a distinct boundary of temperature
and climate between Antarctic and sub-Antarctic zones. It is a fairly
steady one, shifting less than 100 miles north and south from year
to year. Across the boundary the temperature differential appears
to be slight—only 7.5 to 11 degrees Fahrenheit in summer and 2 to
5.4 degrees in winter. But the effect on terrestrial life is quite
remarkable.

South of the convergence the vegetation becomes ever more sparse.
Except for mosses and lichens, it vanishes on the continental mainland. Two species of flowering plants, a grass (*Deschampsia*) and
224

a pink (*Colobanthus*), have been found in the northern reaches of the Panhandle, which, as we have seen, is more closely allied to Patagonia than to Antarctica. So far as terrestrial life is concerned, the convergence is its southern limit, except for aquatic birds and mammals which are nourished by the rich broth of the sea.

The convergence can be sensed by mariners without the aid of a thermometer "as the line to the north of which we felt one day genial air again and soft rain like English rain . . . it was like passing at one step from winter into spring."[1]

On the boundary of the convergence, Kerguelen Island presents a polar landscape of huge glaciers sliding down the mountains into the sea. Five degrees to the south, but north of a dip in the convergence, Macquarie Island presents a more temperate mien, with rich vegetation and a population of wild dogs and cats brought there from Europe.

The convergence runs between the Panhandle and Tierra del Fuego, south of Australia and New Zealand, an invisible but definite frontier of the great ice age in the south. Oceanographically, it is the boundary of Antarctica. According to the Russian scientist V. K. Buinitsky:

The January isotherm of 10 degrees centigrade shows there is no place in the Antarctic region where the climate is warmer than the [Arctic] tundra. Consequently, nowhere can there be any flowering plants despite the fact that some of the northern Antarctic islands are closer to the equator than Moscow is.[2]

Similar comparisons have occurred to others. On his world cruise in 1958 aboard the royal yacht *Britannia,* Prince Philip of England remarked, "It is difficult to believe that South Georgia with its snow-covered mountains, treeless landscape, and huge icebergs floating in its bays is the same distance south of the equator as Manchester [England] is north."[3]

The failure of land forms of life to exist in Antarctica, however, has not intrigued the biologist so much as the profusion of life in the Antarctic seas. The late Carl R. Eklund, one of the American pioneers in Antarctic biology, said that the biota of the south polar regions may be low in numbers of species but it is exceedingly high in numbers of individuals. The Antarctic Ocean is the home of the great whales, of seals without number, and of that flightless social bird the penguin, which has occupied the Antarctic coasts for ages and is the region's true aborigine.

Life in the polar seas, in contrast to the barrenness and sterility of the land, is rich—perhaps the richest of any marine environment. Not only are these waters rich in nutrient salts, but their low temperature enables them to hold more dissolved oxygen and carbon dioxide than the northern seas.

The ultimate food, on which the flourishing Antarctic marine biota are based, is the plankton, the drifting organisms which range from microscopic plants called diatoms, to jelly-like creatures a foot or eighteen inches in diameter. For the biologist, the phytoplankton, which are plants, are basic to the ocean food chain, supporting the zooplankton and all other marine animals. The tiny, microscopic-sized phytoplankton have the ability to conduct photosynthesis and can absorb minerals, nitrates, and phosphates directly from seawater. Phytoplankton vegetation is luxuriant in the seas south of the convergence. The billions upon billions of one-celled diatoms stain the ocean a shoe-polish brown between the ice floes.

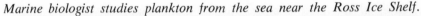

Marine biologist studies plankton from the sea near the Ross Ice Shelf.

Feeding on the phytoplankton are the shrimplike, reddish crustaceans called krill, which have a transparent shell. In Antarctic waters, the red krill (*Euphausia superba*), grow to three inches in length. Swarms of them literally color the sea red. Krill are the principal food of the gigantic blue and finback whales. The abundance of these whales therefore depends on the krill, which, in turn, depend on the phytoplankton that depend on the dissolved minerals, salts, and gases in the sea.

The krill form the key link in the food chain. Squids, which make up the food supply of seabirds, seals, and the sperm whale, feed on the krill. The nine-foot crabeater seal, the emperor penguin, and the petrels dine directly on krill and other crustacea.

But except for the cries of the seabirds in the green-blue skies, the ice cap is silent, dead, in contrast to the noisy, churning life of the ocean. Only man can live on the land, and he must bring his shelter and food with him, unless he turns, as the seals and penguins have, to the sea for food.

During the 1963–1964 austral summer botanists from Ohio State University planted Kentucky bluegrass seed in a small plot of bare ground at Hallett Station on Cape Adare, one of the northernmost points of the Ross Sea region. The seeds were sown on November 11, and by New Year's Day the first sprouts had appeared. The grass continued to grow until January 11, when it began to die.

The experiment was made as a follow-up to work by Argentine botanists, who had succeeded in growing bluegrass on the northern tip of the Panhandle. This area is also south of the convergence, which flows through the Drake Passage between the Panhandle and Tierra del Fuego.

But, as in the experiment at Cape Adare, the Argentine experimenters found that the grass grew for a while and then withered away. None of the sprouts at Cape Adare developed a second leaf—the indication that grass is thriving.

Lichens remain the only land plants in the continental region. They grow so slowly that their spread over the exposed rocks and nunataks can be measured in centuries. They are, of course, a symbiosis of alga and fungus, the alga providing the food by photosynthesis and the fungus giving the plant its water and support on the rocks.

Biologists have speculated that in the event of a warming of climate, which would provide more moisture, lichen growth would be accelerated and that, as the snows retreated, the lichen would spread, the forerunner of more elaborate vegetation. Millions of tiny seeds and spores must be blown into the region by the winds, but under present conditions they do not have a chance.

Birds that walk like men

One of the most highly evolved products of the plankton pastures of the sea is the penguin, a race of aquatic, wingless birds. They have become the trademark of the Antarctic since Von Bellingshausen and Dumont d'Urville first saw them on the coasts.

The penguin has adapted physiologically and behaviorally to the ice age. The wings have become flippers, covered with small, scaly feathers. Paleontological evidence shows that the penguin was native to Antarctica before the ice age. There is reason to suppose that the bird that walks like a man originated on the Antarctic land mass and spread throughout the southern hemisphere from there. Penguins do not live in the northern hemisphere, but their distant relatives, the great auks, did until they were exterminated in the nineteenth century.*

Fossils found by the Swedish South Polar Expedition of 1901–1903 on Seymour Island, off the Panhandle coast, show that there were seven genera of penguins in Eocene times, 60,000,000 years ago. Similar fossils have been found in New Zealand, Australia, and Patagonia, showing hemispheric distribution of the birds. In comparing the fossil creatures with modern penguins, Eklund concluded there has been little change in the hard part of the bird since Tertiary times. The flippers evolved quite early, for no fossils of winged penguins have ever been reported.

One of the interesting features of the Seymour Island fossils is evidence that some species of penguins reached a height of 5 feet. No Antarctic penguin is that tall today. The tallest, the emperor, ranges from 38 to 40 inches in height and weighs up to 90 pounds. The king penguin is somewhat smaller. The scrappy little Adélie penguin, the clown of the whole tribe, ranges from 16 to 20 inches tall and weighs 9 to 13 pounds.

The present and past distribution of penguins throughout the southern hemisphere suggests some kind of land connection between Africa, Antarctica, South America, Australia, and New Zealand, and hence may be cited as further evidence for Gondwanaland. It seems unlikely that the bird, though an excellent swimmer, could have tra-

* The term "penguin" is probably derived from the Latin *pinguis,* meaning fat. It was first applied to the great auks of the Atlantic islands, *Pinguinus impennis,* and then transferred to the non-flying species, south of the equator, for which it became distinctive after the extinction of the great auks in 1845.

Emperor penguins and chicks at Cape Crozier, eastern end of Ross Island.

versed the vast distances across the Pacific Ocean between Antarctica, Australia, New Zealand, and Africa, even by island-hopping.

In common with other seabirds, the penguin is able to remove salt and other minerals from sea water with marvelous efficiency through a gland behind the nose. The gland is reported to be ten times more efficient for this purpose than the kidneys.

Like other birds, the penguin has a homing "instinct" which enables it to return to its native rookery year after year from the northern limits of the sea ice, where it is driven in search of food during the fall and winter. The homing pattern is highly developed in the Adélie. Five male Adélies were banded by Richard L. Penney, a zoologist then at the University of Wisconsin, and flown by aircraft from their native rookery on the coast of Wilkes Land across the ice cap to McMurdo Sound, where the birds were released. The date of the "kidnaping" was December 3, 1959. The following October, Penney found three of the five back at their nesting grounds in Wilkes Land. They had apparently waddled, bellyslid, and swum 2400 miles around the coast, at the rate of about 8 miles a day. Penney reported that he had moved Adélies 35 miles inland from their nests on several occasions and that the birds walked or tobogganed "home" every time.

Everyone who sees penguins is struck by their mien, which is a caricature of that of a pudgy, pompous man, and their gregariousness and highly structured behavior toward one another enhance the image. The coastal rookeries are black with thousands of birds, standing about in groups, nodding, bowing, saluting, fighting, or simply waddling around. G. Murray Levick, the zoologist on Scott's *Terra Nova* expedition, described the Adélies at Cape Royds as "smart little men in evening-dress suits with a shimmering white front and black back and shoulders."

There have been eyewitness reports of Adélies moving in unison, marching first in one direction, then in another, in response to cries from a leader. I have seen groups of them move in swirls. Other observers say they have seen the birds "drill." The motivation of such coordinated flock behavior is obscure.

During this century five species or subspecies of Antarctic penguins (family Spheniscidae) have been identified: Adélie, chinstrap, gentoo, king, and emperor. However, there are probably only two true species on the ice cap, the emperor (*Aptenodytes forsteri*) and the Adélie (*Pygoscelis adeliae*).[4] Chinstraps and gentoos may be classified with Adélies, and the king penguins with emperors.

The Adélie is the most amusing and most carefully studied of the

Adélie penguins at a large rookery on Cape Adare.

penguins. A surprising amount of the Adélie's group and individual behavior is suggestive of that in primitive human groups. Courtship and mating patterns are clearly discernible, even to untrained observers. During the breeding season penguins are monogamous, for environmental reasons. They identify and care for their young until the chicks are fledged and able to fend for themselves in the sea. Family relationships are evident. The hungry chick comes squawking to its parents.

Unattached or "unemployed" males, who are not occupied by sitting on eggs, courting, or nest-building, may perform guard duties or help collect food. Interpersonal relationships on both hostile and cooperative planes are apparent.

The private life of penguins

L. E. Richdale, a New Zealand zoologist of the University of Otago, describes the sexual behavior of the yellow-eyed penguins of New Zealand as follows.

The male penguin stands 6 to 15 feet away from a female and appears to be indifferent to her. Suddenly, he turns around and, with neck arched, beak nearly touching the ground, and flippers extended stiffly forward, he walks rapidly up to and past her with mincing little steps. He then halts, with his back to her, his neck outstretched, and flippers still extended. He poses in this position for about five seconds. Then he lowers his head and peers at her with one eye over his shoulder, as though to note the effect of the demonstration. Richdale calls this pattern the "salute."

Richdale describes another mannerism that he calls the "throb." Everyone who has observed penguins for some time has seen it. The penguin's skin and feathers begin to vibrate or pulsate rapidly, accompanied by a throbbing sound. The throb is associated with an emotional state in the bird before or during courtship, or during mating.

Like people, penguins, according to Richdale, follow the courtship patterns of keeping company, pairing off, and even divorcing. Now and then one of the mates will leave the other and take up housekeeping with a third bird.[5]

There is a widely believed "legend" among Antarctic personnel that penguins are unable to differentiate between the sexes. The gentoo, which is slightly larger than the Adélie, is reputed to attempt to deter-

Left to right: *An Adélie penguin sits on her nest of pebbles at the large Cape Royds rookery on Ross Island; nearby two chicks feed from an adult, back from a food-gathering expedition to the sea; a young wanderer is*

mine which of a pair will be the mother penguin by a test of strength. The weaker of two contending birds pushing against each other is the female.

Prince Philip found the allegation of sexual ignorance among penguins "rather depressing" to believe. "However," he wrote, "there is no doubt that courtship consists of offering small, flat stones for the nest."[6] According to an interpretation popular in the McMurdo Sound region, the witless male penguin wanders around the rookery with a stone in his beak and offers it to every other penguin he sees. The one who takes it from him in its beak is a female penguin.

Richdale's yellow-eyed penguins of New Zealand are a highly sophisticated group, proving that such reports about the sexual behavior of penguins are quite libelous. Males and females have no trouble telling each other apart, and several males will pursue the same female.

In his natural-history report of the *Terra Nova* expedition, Levick has contributed his observations of the boy-meets-girl problem among the little "people."[7] As the nesting season arrives, he wrote, the female scrapes a small depression in the barren soil of the rookery and sits down in it. Lonely and forlorn, she waits until a male approaches her with a stone in his beak. She remains indifferent until he deposits

scolded by an adult guarding its own nest; while a stray chick shivers, away from its nest, a prized pebble is stolen by an unemployed male; at a neighboring nest a chick is warmed and groomed by its parent.

the stone on the ground and waits. If she picks it up and adds it to the "nest" she has accepted him as a mate.

After the mating, the male brings more stones to the nest. An unemployed male may show up and become a rival for the female. He makes this clear by pantomiming the ritual of picking up the stone and offering it to the female for the nest. His efforts to break up the new marriage are usually ignored by both parties, and in that case he continues to waddle on down the line to break up another happy home if he can. Sometimes he is assaulted by the jealous husband.

Penguin cocks frequently fight over a hen or a nest site. While they don't fight to the death, they go at it hard enough to wound each other with their tough, horny beaks. Penguins which have lost one eye or with recently dried blood on their feathers appear fairly frequently. Observers have related instances when cock fights have been broken up by other penguins when the battle went on too long.

The "bugged" egg

During the Adélie nesting season in the austral spring the hen lays two eggs within three or four days of each other. Then she and her mate take turns incubating them by holding them between their feet and against their bellies for thirty-four to thirty-seven days.

In 1958 Eklund devised a unique telemetry experiment at Cape Adare which revealed for the first time that penguins keep their eggs at a temperature of 92.7 degrees Fahrenheit, irrespective of the air temperature. He robbed a nest of one egg, sawed it in half, emptied it, and inserted a small telemetry radio that would broadcast the temperature of the shell. Then he glued the two halves of the eggshell back together and slipped the "bugged" egg back in the nest. The penguin parents accepted it and kept it warm during the remainder of the incubation period.

The earliest description of penguin nesting was given by Von Bellingshausen, who observed emperors: "We saw each bird holding an egg between its feet, pressing its nose to the lower part of its stomach in which the egg makes a small depression. The underside of the egg lies on the feet and thus is held strongly. In order not to drop the egg, penguins will jump with both feet when moving it."[8]

The emperor lays a single egg in midwinter, about July 15. The egg is incubated while the penguin is on sea ice during winter darkness, when the temperature may fall as low as 50 degrees below zero. The incubation period is about 50 days, during which time the internal temperature of the egg is kept high by the parents. When the eggs hatch in the spring, the helpless chicks are held by one parent while the other gathers food. Later, as the chick matures, other penguins in the colony may look after it while both parents are searching for food. The chick is finally fledged in February.

Because their eggs are not laid until spring, the Adélies have much more to fear from the carnivorous skuas, big migrant birds which fly into Antarctic coastal regions in the spring. The skua's favorite food is penguin eggs, which the big bird spears by swooping down on an unprotected nest and impaling the unguarded egg on its beak. The skuas also devour penguin chicks.

An adult Adélie is more than a match for a skua. In fact the Adélie will even attack a man when it believes its nest is threatened. Signs of a penguin's impending assault are unmistakable. The penguin begins to charge and retreat in pantomime, as though to frighten off its enemy by threat. It waves its flippers and squawks like an enraged gander. Finally it makes a blind rush, pecking away with its horny beak and thrashing with its flippers. After a few moments, the penguin scoots back on its tail and flees wildly, running in a wide circle which brings it back once more to confront its enemy.

Savaged by a penguin

One November afternoon in 1959 at Cape Royds I was charged by a furious Adélie, which detached itself from a conference around a nest. Presumably it was an unemployed male guard.

The little fellow strutted toward me, observing me first with his left eye and then with his right. Penguins have their eyes on the sides of their heads, and they skitter and yaw from side to side as they "charge" their adversary. As he made several blind rushes which halted within two or three feet of me he attempted to turn both eyes inward toward his beak, as though sighting along it.

Suddenly, waving his flippers and lowering his head, he charged beak-first into my heavy rubber thermal boots. The impact was surprisingly hard, and I reached down and swatted him across the side of the beak with the palm of my gloved hand. The beak felt as hard as a piece of steel.

My assailant scooted away, circled me, keeping me fixed with one eye and then the other, until he came close enough for another lunge. This time he pulled back before I could swat him again.

The third time he charged I realized he would continue to attack until he was driven off. I retreated a few steps, aiming a camera at him, tripped over a rock, and sat down on the sharp, porous volcanic rubble. Nearby penguins all turned to stare, and my assailant hurriedly backed away. He was satisfied. He dropped his head and seemed to sag in exhaustion. In this mood of weary victory, he allowed himself to be picked up and photographed—the winnah!*

The next austral summer, when I returned to Cape Royds, I looked for my old enemy. One of the birds seemed to be stalking me for a time from a distance, but I could not decide whether it was he or not. To people, penguins look alike.

Footprints in the snow

While Adélies are the most thoroughly studied birds in Antarctica, their behavior can be very mysterious. During the IGY, as I have related earlier, evidence of penguins' ranging far inland was found by two traverses.

On December 31, 1957, the Ellsworth traverse found the track of an emperor 240 miles from the seaward edge of the Filchner Ice Shelf and 300 miles from the nearest known emperor rookery. Two days later the Byrd Station Traverse, 400 miles away, sighted footprints, belly-slide marks, and droppings of an Adélie penguin 186 miles from the nearest coast.

Neither of the penguins was sighted. Both were headed away from the coasts, not toward them. No explanation has been offered as to what the birds might have been doing so far from a source of food, or where they were going.

Penguin locomotion on land or ice does not inspire any confidence in the ability of the birds to travel very far or very fast on the ice cap. They walk and toboggan on their bellies, propelling themselves forward with their feet and balancing with their flippers. In the water their movements are swift and graceful, in contrast to their frantic clumsiness on land or ice. The Adélie is such a powerful swimmer that it can leap 6 to 8 feet out of the water to clamber up on an ice floe.

Bipolar birds

In addition to the aboriginal penguin, Antarctica is the seasonal home of twenty species of birds. On South Georgia Island birds like the

---

* The episode made a merry little item in a New Zealand newspaper under the heading: "U.S. Journalist Savaged by a Penguin!"

pipet, the lark, the teal, and the dovelike sheathbill appear in the summer, but they are not seen on the continental ice cap.

The biggest bird of the southern ocean is the wandering albatross, which is white except for black on wingtips and tail. With a wingspan of 9 to 11 feet, this huge seafaring soarer can cover 300 miles a day "with no sweat," according to Navy observers. The noble bird, immortalized in Coleridge's *Rime of the Ancient Mariner,* is, alas, a scavenger of ships' garbage, like most of the other birds of the polar seas.

Sailing down to Antarctica from New Zealand, one sees at least three species of petrel. The cape pigeon, one of the most common, has a three-foot wingspread with black and white "polar" plumage. Its first cousin, the Antarctic petrel, has brown plumage on the upper body and white underneath. The giant petrel, with a seven-foot wingspan, is gray and brown—the harder to see as he crouches in the rocks of a forlorn coast, awaiting a chance to dive into a penguin nest and snatch an egg or a newly hatched chick.

Over the crags of the Trans-Antarctic Mountains sail the skuas. Perhaps the largest long-range flier of the Antarctic is the south polar skua, the mortal enemy of the penguin. In the spring the skuas migrate to Antarctica from islands in the North Atlantic Ocean. They range far inland and have been seen eight hundred miles from the coast.

Like penguins, the skuas are monogamous. The necessity of keeping their eggs protected against the cold winds has forced them to

A female skua sits on an unhatched egg in her nest at Cape Evans, while a young chick nestles close to her for warmth.

evolve a family life. At Cape Royds, brown and white against the stark rocks, they perch like huge falcons in the upper crags, staring at the penguins on the beach below. For their part, penguin guards stare upward at the skuas.

Now and then parents get careless. They may wander away from the nest, leaving the eggs exposed for just a few seconds. That is all it takes. Down swoops the skua with the speed of a jet airplane. It spears an egg with its beak and rises swiftly and away. Pathetically, the hapless parents waddle after it, emitting strangely human cries.

The long-distance flying record is held by the arctic tern, which was first seen summering in the Weddell Sea on the Atlantic side of Antarctica in 1904 by naturalists of Bruce's Scottish National Expedition. The tern flies a round trip 22,000 miles long, almost from pole to pole, to stay with the sun. The bird must average 20 hours of daylight all its life. As soon as the sun begins to dip in Antarctica and the 24-hour daylight fades, the tern begins heading northward.

For the most part the winged birds of Antarctica remain aloof from man. They are visible soaring high in the sunlit heavens, or roosting on the volcanic crags of Ross Island, or flying low over the sea, fishing. Theirs are the only cries that break the silence of this land over most of its soundless reaches.

The fur-seal massacre

Seal-hunting and exploration are inextricably tangled in the early history of man's penetration of Antarctica. Seals have been known in southern waters since the time of Sir Francis Drake.

The first men to see the ice-clad mountains of the Antarctic Panhandle were sealers—Nat Palmer from Stonington, Connecticut, and William Smith from Liverpool, England. Hope of profit rather than the stirrings of intellectual curiosity brought the sealers to Antarctica, and profit there was. So intensively were the fur seals hunted on the beaches of the sub-Antarctic, and later the Antarctic, islands that by 1830 they were nearly all wiped out. Only the most stringent conservation measures have saved them from extinction.

From an economic viewpoint, the seals are of two kinds: those with marketable fur and those without it. Most of the fur seals belong to a group called the eared seals, because their ears are visible. There are two other groups—the walruses, which are not found in the Antarctic, and the true seals, which have a wide Antarctic distribution. The only true seal which is hunted for its fur and also its oil is the big elephant seal. The others have shaggy coats of unattractive hair which have turned out to be their salvation.

The fur seals lent themselves to extinction. Not only did their fur make them a target for hunters, but they are the easiest of all animals to hunt because of their mating habits. The eared seals are polygamous. The bulls maintain large harems on the stony, barren beaches of the sub Antarctic and Antarctic islands, to which the population invariably returns year after year.

True seals, on the other hand, spend most of their time in the water. Their mating is more casual. They do not cluster on the beaches like their furred relatives. Consequently they are harder to find, harder to kill if anyone wants their hides. The exception is the elephant seal which, though a true seal with ears tucked away out of sight, is also a harem-gathering animal and hence prefers the beaches.

Most of the fur seals have, beneath an outer coat of long hair, an inner one of short, dense fur, which became highly prized in Europe and China for coats early in the eighteenth century. In 1750 Chinese furriers invented a method of getting rid of the longer hairs without damaging the soft undercoat. The price of sealskin skyrocketed on the Chinese market.

Weddell seals basking in the austral summer sun on the Antarctic ice.

In 1798, for example, more than a million sealskins, procured off the coast of South America, were sold in Canton. At the beginning of the nineteenth century European furriers adopted the Chinese technique of processing sealskin, and there was another boom in Europe. That nearly wiped the fur seal off the face of the southern hemisphere.

As ships and navigation improved, sealers ranged ever farther southward, now in search of the big, stupid elephant seal, the largest and most profitable of all. The bulls, fifteen to twenty feet long and weighing three to five tons, were a rich source not only of fur but of oil finer than whale oil.

One sees no fur seals today in Antarctica, although they are reportedly coming back in considerable numbers on some of the sub-Antarctic islands.

Contented true seals

The seals most frequently seen on the coasts of West Antarctica and the Ross Sea embayment are the Weddell seals, a species of true seals first described by Captain James Weddell. They appear to be the most numerous seals in the Antarctic, although no one has ever taken a census.

One finds them on the northern rim of the Ross Ice Shelf, sunning themselves or popping up through holes in the ice with large, struggling fish in their mouths. Slaty gray with a mottling of white and silver on the belly, the Weddell seal is an unambitious, somnolent animal. It grows 9 or 10 feet long and weighs up to 1000 pounds.

Until the advent of man, the seals had no natural enemies except the killer whale and the carnivorous leopard seal, which occasionally attacks its fat cousins, who are not conditioned to fight or flee other animals. It is quite easy to approach them on the ice and even roll the pups over. Adventurous visitors try to "ride" a seal. This is not recommended. An alarmed seal can do a good deal of damage to a human being with a mere flick of its powerful tail.

During the austral summer of 1960 an 800-pound Weddell female, apparently seeking a mate, attempted to wriggle into a field tent—attracted, witnesses believed, by the lusty snores of a sleeping geologist. The witnesses said they frightened the seal away. Their colleague slept through the entire episode and doesn't believe a word of it to this day.

Another common variety of true seal is the crabeater. It lives mostly in the open sea between the ice cap and the pack ice. Crabeaters,

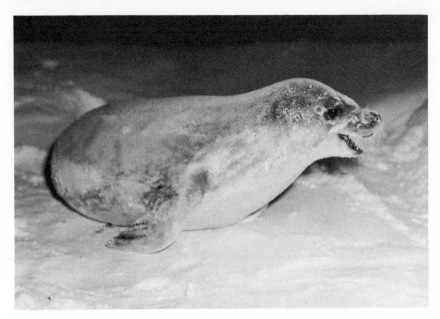

A crabeater visits Cape Evans, though this species is rarely found there.

which dine on the red krill, grow to 7 feet in length. Their coats are dirty brown or stained yellow, molting to white with a reddish tinge.

In contrast to these peaceful animals, the leopard seal is a dangerous carnivore. It is big, powerful, and agile. It will leap out onto the ice to pursue its quarry, including man. The bulls weigh up to 1200 pounds and attain up to 12 feet in length. With its small, snaky head and streamlined body, the leopard seal is readily identifiable. Its slaty-gray hide is splashed with bright yellow and black. The animal seems to flash in the water.

The leopard seal feeds mostly on penguins. It seizes the Adélie, skins it alive, spits out the skin and feathers, and gulps down the bird in seconds. It will attack and butcher other seals if hungry enough. Its mortal enemy is the killer whale.

Least well known of the true seals of Antarctica is the Ross seal, a clumsy animal with a gray-brown coat. Like the crabeater, this one lives in the open sea, but apparently frequents regions not often visited by man. Less than a hundred Ross seals have been reported.

Since the Antarctic seal, among the animals which bear their young alive and suckle them, dwells in the coldest climate, there has been a great deal of curiosity about its physiological adaptation to the environment.

In 1963 six Weddell seals were airlifted from McMurdo Sound to the New York Aquarium for study. In the following year it was not certain that all of them would survive. Of particular interest to the associate curator, Carleton Ray, and his associates is the seals' heat-regulating mechanism.

Seals maintain a body temperature about the same as that of human beings. They can dive to great depths in the sea, making adjustments for extreme pressure which enable them to zoom quickly up to the surface without getting the "bends" as human beings would. At Mc-Murdo Sound depth gauges were attached to seals being studied early in 1964. Several Weddell seals dove to a depth of 1480 feet in the sound for bottom-dwelling fish and surfaced with the fish in their mouths. At that depth the water pressure was nearly 700 pounds per square inch—50 times the atmospheric pressure at the surface.

However, it is the sperm whale, not the seal, which holds the record for deep-sea diving among mammals. Whales can remain submerged about 45 minutes, whereas the Weddell seals at McMurdo Sound stayed under water less than 30 minutes. The longest dive by a seal while the dives were being measured was 28 minutes.

In the opinion of a number of marine biologists the Weddell seal is the most highly evolved of all the seal species and the most successfully adapted to its environment. If well-nourished contentment is the standard of successful adaptation, the Weddell seal has certainly arrived at a harmonious adjustment to its environment. There it lies in the summer sunshine on the hard ice, while the breezes whip about it at 25 miles an hour and 17 degrees below zero Fahrenheit. It is so content it can even tolerate pesty biologists trying to analyze it.

Seals are the top of the Antarctic food chain, or were until man came and started killing them for fur or butchering them for sled-dog food. Now and then, the leopard seal or the killer whale eats one, but seals have no natural enemies bent on their liquidation except man. In that respect the seals have something in common with the great whales.

Thar she blows!

The seas around Antarctica are the home of whales. Nearly 90 per cent of the modern whaling industry is concentrated there. Aside from its scientific lure, Antarctica's main attraction in the nineteenth century was the seal and then the whale. But the fur seals were nearly gone when whaling began in the southern ocean—and that was not long ago. It was not until 1893 that the first whale was killed south of 55 degrees south latitude.[9]

Whaling on a regular basis in Antarctic waters commenced in 1904,

Above: *A beatific seal with her young pup in the water.* Right: *A Weddell seal on the bay ice is unconcerned at the approach of McMurdo personnel.* Below: *A member of the McMurdo biology laboratory takes a milk sample from a Weddell seal for a study of mammal physiology.*

when an Argentine firm set up a whaling station on South Georgia. During the austral summer of 1904–1905, the whaling catch totaled 195 whales. That was the beginning.

Antarctic whaling may be the end of the whaling industry, which began more than a thousand years ago, when Spanish and French Basques hunted the big cetaceans in the Bay of Biscay. With primitive boats and lances they were able to kill the slow-moving right whales, which after death floated on the surface, instead of sinking to the bottom. Once they had killed a whale, these medieval hunters dragged it ashore and cut it up.

Within four hundred years the Biscay whalers had ranged as far as Iceland, and no doubt some of them sighted the coasts of Greenland and Newfoundland before Columbus sailed, unaware that they were seeing a New World.

With the discovery of Spitsbergen by the Dutch at the end of the sixteenth century, whaling moved deep into the Arctic. The Dutch became dominant in the industry. Between 1679 and 1688 it was reported that they killed an average of a thousand whales a season.[10] British firms entered whaling on a large scale in the eighteenth century.

In the year 1712 a party of Nantucket fishermen managed to harpoon a sperm whale and found that its oil was higher in quality than that of the right whales, to which hunting had been limited for more than seven hundred years. Moreover, they discovered a waxy substance called spermaceti in the head. It made the finest candles in the world. Soon this whale wax was in great demand.

Between 1804 and 1876 more than 400,000 right and sperm whales were taken by Americans. The kill rate was higher than the whale-population replacement. Whaling soon declined in the northern hemisphere for the balance of the nineteenth century.

Accounts by Ross and Wilkes of great whales plowing through the Antarctic seas drew the attention of whalers to the south. Reports of the enormous size of the blue whale excited whaling men from New England to Norway. In his ship, the *Antarctic,* the Norwegian explorer Otto Nordenskjold took careful note of whales in the southern ocean. His captain, Carl Anton Larsen, essayed to hunt the blue whale with right-whale weapons, but failed to hold the monstrous animal. More powerful weapons, bigger ships were needed.

Seeing the possibilities of southern-hemisphere whaling, Larsen aroused the interest of a group of Argentine investors, who financed the first whaling station in the Antarctic—the Compania Argentina de Pesca Station, which Larsen built in 1904 at Grytviken, South Georgia. The Norwegian sea captain obtained a twenty-one-year lease from New Zealand to operate in the seas between that island country and Antarctica.

Larsen developed the modern whaling ship, where whales can be processed on deck. Huge whale-factory ships began appearing in Antarctic waters after World War I. The catches were enormous. They flooded the market in 1930–1931, coincident with the depression. The bottom fell out of the whale-oil market in those years. The entire Norwegian whaling fleet was laid up for an entire year.

Unrestricted hunting depopulated the southern ocean of whales, as it had the northern seas. The naturalist Robert Cushman Murphy of the American Museum of Natural History recalled that when he visited South Georgia in 1912–1913 the humpback whale comprised 98 per cent of the catch. But 20 years later, he said, the species had become so depleted that it made up only 2 per cent.[11]

Abandoned during World War II, whaling was resumed in the southern ocean after the war. By 1951 there were ten thousand whaling men in Antarctic waters, about seven thousand of them Norwegians. The Russians had their huge whale-factory ship, the *Slava,* there. Norway operated ten factory ships; the British had three, the Japanese two, and the South Africans and Dutch one each.

During 1953–1954 about 388,600 tons of whale oil were produced from the Antarctic catch, equal to the fat of 80,000,000 sheep.[12] In Europe and Japan, whale oil is used for margarine and synthetic lard. The poorer oil is used for machine lubricants and tanning. Spermaceti is used for soap, beauty creams, ointments, and decorative candles. Fancy birthday candles are made of it, for spermaceti gives off a particularly gay, sparkling light.

Since 1945 whaling in the Antarctic has been governed by international convention, which limits the hunting season to 4 months in a year. In the 1962–1963 season the convention prohibited the killing of more than 15,000 blue whales, or their equivalent in other species, each year by all nations. One blue whale equals 2 finbacks, 2.5 humpbacks, or 6 sei whales.

The blue whale is the largest of the cetaceans and is apparently the most gigantic animal that ever has existed on this planet! Some are 108 feet long and weigh 160 tons and may yield 20 to 50 tons of oil and edible meat. Such a whale equals the combined weight of 25 elephants or 150 bulls. Its vertebrae are the size of big bass drums. Some of its long bones are as big as steel girders. Compared to the

A Stanford University zoologist observes sei whales feeding in a channel cut into the frozen bay waters of McMurdo Sound by icebreakers.

modern blue whale, the giant reptiles of the Mesozoic Era were pygmies.*

Next in size to the blue whale is the finback, or fin whale, averaging 65 feet in length and weighing 50 to 60 tons. Some finbacks have been reported recently to be as much as 90 feet long. After the finback comes the sei whale, which rarely exceeds 65 feet. The humpback is slightly smaller.

These whales are called "baleen whales" (suborder Mysticeti) because, instead of teeth, bony ridges in the roofs of their mouths and upper jaws have become enlarged into baleen, a fringe of long, horny plates. When the whale gulps a mouthful of water, the baleen acts as a sieve, holding the shrimplike krill but allowing the sea water to drain out. Baleen, known also as whalebone, was used extensively in the nineteenth century for women's corset stays and for umbrellas. Some is still used by European milliners and couturiers in lieu of the less expensive steel, which has largely replaced baleen.

* *Brontosaurus* was 67 feet long and weighed perhaps 35 tons; *Diplodocus* reached 87 feet, weighing 25 tons. One of the heaviest of the giant reptiles, *Brachiosaurus,* weighed 50 tons. He was so fat he could hardly walk and had to wallow in a swamp.

Among the toothed whales, only the sperm, or cachalot, has been found commercially valuable. The cachalot yields 8 to 20 tons of oil, depending on its size, which ranges from 32 to 65 feet. Cachalots are easy to identify. Any amateur can spot one at a glance. One-third of the animal's length is its enormous head. Melville's Moby Dick was an albino sperm whale.

The toothed whales (suborder Odontoceti) are more numerous but generally smaller than the baleen whales. Odontoceti includes sixty-five species in four families: sperm; bottlenose, or beaked, whales; marine dolphins and porpoises; and the fresh-water dolphins.

Among the mammals of the sea, or on land, for that matter, the dolphins appear to be distinguished by their intelligence. The smaller dolphins respond to man quite well, but not the larger ones, which are commonly called whales—the killer whale, white whale, and pilot whale—though these have dolphin characteristics, such as teeth in both upper and lower jaws and a dorsal fin halfway along the back.

Killer whales are sometimes seen in Antarctic waters. They have attempted to attack men, dogs, or ponies on ice floes and have demonstrated the ability to work as a team to get at their prey. There is no mistaking the killer whale. The dorsal fin is high and quite visible, extending six feet from the back of the whale to its tip.

From the remains of a sperm whale whose lower jaw had become caught in a cable at 540 fathoms, it is evident these animals can descend to 3200 feet. At such depths they hunt squid, remaining under water up to 45 minutes before they surface for the next spout. Some sperm whales bear large sucker marks as evidence of underwater battles with giant squid. Remains of tentacles taken from the stomachs of large bull whales could only have come from squid larger than the fifty-foot specimens frequently found on New Zealand beaches.

How long do whales live? Not so long as their size might suggest. Age studies have been made on some species. They indicate that a female finback has a life span of about fifty years or less. Humpback whales may live 15 to 30 years.

Before World War II a number of humpbacks were marked so that spotters could track their wanderings. During the winter they leave Antarctica and swim north along the coasts of South America, South Africa, Australia, and New Zealand. One humpback, marked with an aluminum plate off Enderby Land on the Indian Ocean side of Antarctica, was seen later in the season plowing along near Madagascar. In the spring the humpbacks return to Antarctica to feed.

Blue whales are believed to swim due north in winter to bear calves off the coast of Lower California—nearly halfway around the world for the round trip.

In 1963 there were seventeen whaling fleets in Antarctic waters, representing Japan, Norway, Russia, Great Britain, and the Netherlands. The United States has not sent a whaler into the Antarctic since 1940, according to the Department of Commerce. The market for whale products has all but vanished in this country.

Compared with whaling in the days of Moby Dick, the modern industry is about as romantic as a stockyards slaughtering pen. The ships are steel, the harpoons are fired like artillery weapons, and the whales are tracked by radar and helicopters.

Hijacking the seals

There are plenty of fish in Antarctic waters, and few fishermen. Regions like McMurdo Sound ought to be a fisherman's paradise on a sunny summer's day, the temperature hovering around zero degrees Fahrenheit and not a cloud in the sky. The kind of fishing men do in this region, however, is not for sport but for science.

For sixty years biologists have been trying to trap fish with nets

In a study of diseases that attack seals, a virologist, seen at right in a diver's suit, investigated the Weddell seal in its natural habitat.

and traps lowered through holes in the bay ice of the sound. Their success has been indifferent. Most of the large fish seem to dwell at the bottom of the sound, about 1500 feet deep. The biologist wants the big fish. He needs to know how big they grow—and how fast.

There are two ways to acquire fish in this region. One is to bore a hole in the ice, which can be a chore, for sometimes the ice is 30 feet thick, though in the late spring it usually is 5 or 6 feet. The fisherman then erects a tent over the hole to keep out the cool breezes, unfolds his camp stool, tunes his portable short-wave radio, drops his net or trap through the hole, and waits.

An easier way is to shadow a lean and hungry-looking seal to a fishing hole it knows of, watch it when it dives into the water, and then, with the aid of several companions, seize it when it comes up with a fish in its mouth and wrest the fish from it. The really big fish in the Antarctic have been hijacked from seals, not caught, netted, or trapped.

Marine creatures are fruitful and multiply. Offshore waters abound in sea urchins, starfish, crustaceans, worms, and mollusks. Shellfish are believed to be more numerous here than in Arctic waters. There are more sponges at the bottom of McMurdo Sound than in many

Most common animal species, found at the sea bottom of McMurdo Sound, include sponges, several bivalves, two asteroids, and a Trematomus *fish.*

parts of the tropical Indian Ocean. The Antarctic sponges grow long, glassy spikes, or spicules.

In calm weather, jellyfish 2 feet in diameter can be seen floating at or just beneath the surface. Off the northern Antarctic islands and in the sub-Antarctic waters north of the convergence, lobsters are plentiful. They resemble huge crayfish and grow 3 feet long.

Plentiful and of assorted sizes are the soft-bodied mollusks or cephalopods—the squids and octopuses. The smaller squids range in size from 10 inches to 6 feet. The larger reach a length of 46 feet. These big fellows have 10 thick tentacles equipped with suckers and claws like grappling hooks.

A giant squid is nearly a match for a whale. Bruises of squid suckers and claws about the huge heads of sperm whales prove its prowess. It takes a powerful animal, indeed, to damage the heads of cachalots, which have been known to smash propellers and the rudder of a ship with their heads, without showing a scratch.

Most of the fish in Antarctic waters belong to the Nototheniid family, some members of which resemble cod and bullheads. One of the largest, *Notothenia rossi,* about 40 inches long, is found off South Georgia and the South Shetland Islands.

In the McMurdo Sound region marine biologists have observed twelve species of fishes in the families Nototheniidae, Bathydraconidae, Chaenichthyidae, and Zoarcidae. The most numerous are the Nototheniidae, represented in the sound by species of big-headed, slender-bodied fish in the genus *Trematomus.*

One of the largest fish ever seen in Antarctic waters was hijacked from a seal by a gang of Stanford University biologists in 1963. They managed to keep the fish alive for about 10 days in a tank rigged up in the biology laboratory on Ross Island. The fish was 51 inches long and weighed 47 pounds. It was dark gray with wide light gray vertical stripes. When the fish was irritated, the stripes would disappear. It was tentatively identified as *Dissostichus mawsoni.*[13]

Studies of the metabolism of Antarctic fish have been of prime interest to marine biologists. Observations show the metabolic rates of Antarctic fish are higher than those of fish living in warmer waters. The Antarctic fish also are able to convert energy in growth more efficiently than fish in warmer latitudes. It is assumed that metabolic and growth differences are due to the lower temperature of the Antarctic seas. In fact the high metabolism of Antarctic fish is referred to as "cold adaptation" by the veteran Antarctic marine biologist Dr. Donald E. (Curly) Wohlschlag of Stanford University.

A Stanford University zoologist examines a large fish (probably Dissostichus mawsoni) *taken from the mouth of a seal feeding in McMurdo Sound.*

Wohlschlag has studied the metabolism of *Trematomus bernacchi,*
a big-headed fish in McMurdo Sound. There the water is as cold as it
can get without freezing to the bottom. "If the mechanism of metabolism were known," he wonders, "could it be manipulated to raise
the metabolic level? To speed the growth of a temperate species at
winter temperatures by raising its metabolism to the higher levels of
polar species at similar water temperatures?" The implication is an
approach to increasing the population of the temperate ocean fishing
grounds.

Wohlschlag and other biologists have found several significant
aspects of adaptation to cold by marine animals in Antarctic waters.
One species of *Trematomus* has mutated and has no hemoglobin, the
oxygen-carrying compound in the blood. Because of the high concentration of oxygen in the cold Antarctic seas, hemoglobin is not needed
to transport oxygen to the tissues. These fish manage without it. The
price of the mutation is that the species is confined to polar waters.

A similar price has been paid by the cold-adapted red krill. In
studies of polar krill, Dr. Mary A. McWhinnie, biologist, and Phyllis
L. Marciniak, then a graduate student, of De Paul University in Chicago, found that the little creatures, which are vital in the Antarctic
food chain, cannot survive in water warmer than 39 degrees Fahrenheit.

Dr. McWhinnie and Miss Marciniak made studies of the metabolism of krill (*Euphausia superba*) and other marine animals in the Antarctic Ocean on the 1962–1963 cruise of the National Science Foundation research ship *Eltanin*. Collected by trawls in the Bransfield Straits, west of the Panhandle, where the sea-surface temperature averaged just above freezing, the krill were placed in refrigerated aquariums aboard the *Eltanin*. When the temperature of the water was raised only to 39 degrees Fahrenheit, the researchers reported, "the animals became quiescent, lost their glassy transparency and died in 24 hours."

Within a narrow temperature range, from 29.5° F. to 37.5° F., the metabolism of the krill remained relatively unaffected. It was clear they could adapt to temperatures within this range. At 39 degrees Fahrenheit and above, they died, apparently from the heat. Dr. McWhinnie noted a cessation of motor activity, and through the animal's transparent shell she observed a marked increase in heart rate, until the heart stopped and the animal became opaque.

Thus, like the mutated fish that lacks hemoglobin, the krill too are confined. In the Antarctic their life boundary is the convergence, north of which they cannot go.

Antarctic fish story

One of the most curious discoveries involving marine life was made by two University of Michigan geologists, Charles W. M. Swithinbank and David G. Darby. On November 8, 1960, they found the remains of fifty partly decomposed fish on the surface of the Ross Ice Shelf about a mile from the ice front. Strewn among the fish were deep-water clams, snails, glass sponges, and corals. Some of the fish—Nototheniidae—were intact. The largest was 65 inches long. There were detached heads that appeared to have come from even larger fish.

These fish were larger than any seen before in the region. Evidently they were bottom-dwellers. If that was so, how did they get on top of the ice shelf, which in that area was 90 feet thick?

A similar discovery was reported a half-century before by Frank Debenham, a geologist on the *Terra Nova* Expedition of 1910–1913. He described the finding of headless fish, along with well-preserved glass sponges and corals, on the ice shelf. Debenham theorized that the animals were pinned to the bottom of the sound at a time when the keel of the shelf scraped the sea floor. They were then frozen into the ice as new ice formed on the under surface of the shelf.

Over many hundreds of years, the fast-frozen fish were lifted upward by the melting of the shelf's upper surface. The theory rested on the supposition that the shelf was nourished from the bottom, rather than on the top.

It is rather difficult to think of this means of shelf-building, for it snows on the shelf and the huge cake of ice is fed also by the outflow from the polar plateau and Marie Byrd Land. Debenham's theory may account for the manner in which marine life "migrated" upward through the ice, but it is not accepted as the means by which the 196,000-square-mile ice shelf is sustained.

Swithinbank and others suggest that the main body of the shelf has accumulated from snowfall on the surface. As we have seen in Chapter 4, Albert P. Crary has found evidence on his traverses that the shelf is an extrusion of the West Antarctic ice sheet.

It may be that the accretion from the bottom up, suggested by Debenham, occurs only in the McMurdo Sound area, where the shelf is thinner than in other parts of the Ross Sea. In the sound, the shelf

A biologist clad in an anti-exposure suit probes the icy waters of the Ross Sea, searching for specimens of Antarctic invertebrates.

does show evidence of some melting during the summer. The ice cover to the north of it goes out in summer, and the region is probably warmed by the volcanic system, manifest by the snow-capped, smoking cone of Mount Erebus.

Swithinbank and Darby reported that radiocarbon dating of the fish remains in New Zealand indicated they were about 1100 years old. Thus, the upwelling process would appear to take about a millennium.

That, however, was not the end of the story. Three years later, in 1963, a New Zealand glacial geologist, Anthony J. Gow, then with the Cold Regions Research laboratories of the United States Army Corps of Engineers, drilled a hole through the ice near the spot where the frozen fish were found. Gow took the ice core from the hole and examined its crystal structure from top to bottom. He reported that the shelf in that section was entirely fresh-water ice—not sea ice, as the bottom-freezing theory of Debenham and others requires.

Gow also found a layer of fresh water between the bottom of the ice shelf and the sea water. It must be due to shelf-bottom melting, he observed. However, this does not preclude the possibility that sometimes the fresh-water layer on the bottom may freeze and become a part of the shelf.

What is truly astonishing about the remains of fish and sponges found frozen in the shelf is the manner in which much of the organic material has been preserved. Even the fragile glass-sponge spicules have remained intact, the clamshells are still hinged, and the coral remains attached to the rock. Only some of the fish have been marred.

The mystery is as baffling as ever. How did these creatures migrate through 90 feet of ice—straight up from the bottom of the sea?

THE ANTARCTICANS

Strictly speaking, man is the only land animal that has been able to adapt to the Antarctic. His adaptation is technological, not physiological. And he can survive there only by bringing his environment with him.

If this kind of adaptation seems to be flimsy and impermanent, consider the Eskimo of the Arctic. He must be able to create a microenvironment—in the form of clothing and shelter—for his survival. In the Arctic, where there are land animals and access to plants, this can be done with natural materials. But in the lifeless latitudes of the Antarctic interior, all the material for survival must be imported.

We see in the Antarctic the beginnings of human adaptation to hostile environments through technology. It may well be an evolutionary development of incalculable value to the expansion of mankind.

Of course man has "conquered" or learned how to live in other hostile regions, but Antarctica is unique in two respects. It is the coldest and highest of the continents and the most inimical to life. And it is the only one where no evidence of human occupation prior to its sighting in 1819–1820 has ever been detected. Is it possible that the great ice cap conceals evidence of some former human occupation?

We cannot know this until we know how old the glaciation is. Indications that it has been as extensive as it is now, or even more so, since

255

the beginning of the Pleistocene Epoch, seem to rule out the possibility of human habitation. If so, the seventh continent is the only land region where man did not set foot until recent times. It is the last frontier indeed.

Human occupation of Antarctica represents a modern stage in the expansion of the species, anthropologically. Even though expeditions have been visiting the region since 1841, and men have wintered over on the continent since 1899, it is only since 1954 that Antarctica has been inhabited continuously.

In the brief time of occupation, what have we learned about the ability of man to adjust here? Strangely enough, the investigation shows that, at the current level of technology, the relationship of man to man has been more troublesome than that of man to the environment.

The environmental feature which affects human beings most significantly is not the cold, against which technology can provide adequate shelter, but the isolation. From this, there is no relief. Antarctic stations, therefore, are natural laboratories for the study of human behavior in isolation.

All the studies so far show that when men are isolated in small groups from the mainstream of humanity, the psychological stresses rather than the physical ones are of primary importance. This is not revealed in the classic literature of Antarctic exploration, which has emphasized the struggle of man against the physical environment. In modern exploration, with the airplane, the icebreaker ship, and the tractor, the physical challenge of the elements has been greatly reduced. Man now becomes preoccupied with the problems inherent in his nature, and these do not lend themselves to a technological solution.

Perhaps it is fair to summarize the physiological and psychological studies of the Antarctic experience with the statement that human adaptation to the Antarctic environment is not physiological, but behavioral.

It may be that human beings who were born and reared in Antarctica might show some signs of physical adaptation, such as the ability of Alaskan Indians and Eskimos to tolerate more severe cooling of hands and feet than temperate-climate dwellers can.[1] So far, however, individuals have not remained in the Antarctic longer than three years at a stretch, and the average length of "residence" is less than one year.

And yet some men are lured back to the ice year after year, by its mystery, its beauty, and its challenge. They are the Antarcticans.

A New Zealand mountain climber instructs American scientists in crevasse extraction and in climbing techniques before they go into the field.

The crew of a Hercules C-130 aircraft unloads supplies at the field site of a scientific party in the Pensacola Mountains.

Their lives are keyed to the region, professionally, intellectually, emotionally.

Heroes with runny noses

In this magnificent and uncompromising land, man has reached the end of the earth. From here, there is nowhere else to go on this planet.

The Antarcticans sense that life must be asserted here. It cannot draw back before these wastes, but must subdue them if only to demonstrate its viability.

From time to time Antarcticans have spoken to one another across the brief decades they have wandered over the ice, in a special way: they have left notes in rocky cairns.

About 325 miles from the south pole there is a stony ridge called Mount Betty, in the Queen Maud Mountains, between the Strom and Axel Heiberg glaciers. In 1963 a geological party led by Dr. Swithin-

To obtain information necessary for the production of accurate maps, two surveyors, landed by helicopter on a 6000-foot mountain called Mount Crash, are using a theodolite and an electronic distance-measuring device.

bank found a cairn there containing a note left by a geological party from Admiral Byrd's first expedition, on Christmas Day, 1929. It reported: "Here we discovered a cache laid down by Amundsen [1911] on his way northward from the Pole. We have carefully replaced all rocks in the cairn and leave it intact as we found it except for the bit of writing which Amundsen left in a tin can." The note was signed by Gould, then a geologist with Byrd.

No matter how much they are drawn to the mountains and the ice, the Antarcticans cannot, it seems, adjust physically to the brutal climate. Unfortunately for the heroic image, most explorers and scientists respond to the cold with runny noses. You see a researcher attired in parka, overpants, boots, and black beard, attempting to set an instrument on a windy ice sheet, while his nose runs and his eyes water behind his sunglasses. A runny nose is the trademark of the Antarctic frontiersman. It is a normal response to excessively dry, cold air.

The notion that men somehow alter their physiology or metabolism in polar regions so as to "get used to" cold has no scientific support. Men do become accustomed to experiencing cold temperatures, and learn not to fear them. But there is no evidence of physical changes that would indicate a polar adaptation in wintering-over personnel.

Early in the IGY two civilian scientists at Little America V volunteered as guinea pigs to find out how long they could endure the winter temperatures outside the station, on the eastern end of the Ross Ice Shelf not far above sea level. This region is considerably warmer than the high ice plateaus of the interior.

The volunteers dressed in clothing which they considered adequate for winter weather on the shelf and went outside in temperatures ranging from 22 to 49 degrees below zero Fahrenheit. The subjects were instrumented so that their skin and rectal temperatures would be continuously recorded.

Almost immediately after they were exposed to the cold air, their skin temperatures fell and their rectal temperatures rose. It was assumed that this represented an insulation response, due to constriction of the outer blood vessels in an involuntary reaction to cold air.[2]

More recently experiments at Yale University have suggested such responses are dictated by specialized brain cells which act as thermostats insofar as they react to minute changes in blood temperature. These cells are believed to trigger a whole series of physiological events. They increase thyroid secretion, which speeds up metabolism, the rate of fuel-burning, and also stimulate hunger—in anticipation of the body's need for fuel.[3]

When the subjects exercised vigorously, however, the insulative response was reversed. Heat produced by the muscles lowered the rectal temperature and raised skin temperature. Frostbite appeared on the face of one subject, even though he was perspiring from exercise.

The exposures lasted for periods of 40 to 165 minutes and were ended when finger and toe temperatures began to drop. The falling temperatures of the extremities indicated that body heat was being lost rapidly.

Tests also were made that year (1957) to find out if any metabolic changes had occurred in the men at Little America after a year in the Antarctic. No changes were detected. Most of the wintering-over personnel had put on some weight. This was a response not to climate, but to overeating and comparative inactivity during winter confinement.

The polar doctor book

One of the most useful medical results of the Antarctic experience is a handbook on polar health and safety, written by Captain Hedblom, the Navy physician who in 1960 found the only milk in the world uncontaminated by radioactive fallout, at Little America.

Nearly everyone who visited the Antarctic during the early Deep Freeze Operations met and listened to Doc Hedblom. His exposition on polar medicine was given to each new group of arrivals and certainly was the highlight of the round of briefings. Usually billed under the title of "How to Use Your Head to Save Your Ass in the Antarctic," it was a masterpiece of medical wisdom, horse-sense, and polar lore. A newspaperman whom Doc treated for altitude sickness after a flight to the south pole once characterized the broom-bearded physician as having "the bedside manner of a polar bear."

Doc Hedblom's *Polar Manual* is useful in any cold climate. It warns that one of the common afflictions to look out for in the Antarctic is dehydration. Since the region is meteorologically a desert, the air is quite dry and moisture escapes freely from the body, especially in summer, when outer clothing is worn loosely and the sun is shining.

However, even in the summer it is quite cold. And one of the responses to cold, as we have seen, is increased peripheral resistance to the blood flow. The work of the heart is increased; blood pressure and pulse rate rise. In compensation, Doc's manual points out, the

Captain Earland E. Hedblom, physician to the Antarcticans.

urinary output is increased. The result is dehydration. Along with it there may be some temporary concentration of the blood, which contributes to sluggish bowels, hemorrhoids, and headaches.

"Add excessive drinking of strong coffee, tea, and chocolate as increased kidney stimulants, and alarming results of dehydration sometimes occur," the manual states. This is an exceedingly good warning for new arrivals who tend to overdrink coffee—a tendency in polar regions anyway, and a tradition, certainly, at Navy stations.

Another common environmental affliction is snowblindness, or, more properly, sunblindness. Its effects show up from two to twelve or more hours after the victim has been exposed to ultraviolet light from the sun. The light is seldom direct. It is usually reflected from the snow surface and may not be noticeable while it is inflicting a burn on the conjunctiva of the eye, the mucous membrane which lines the inside of the eyelids.

Symptoms are extreme pain in the eyes, caused by swelling of the conjunctiva. So intense may be the pain that the victim is sometimes afraid to open his eyes to any kind of light. If he is on the trail, he becomes helpless. The eyes keep filling with tears and the eyelids smart and feel scratchy.

The symptoms last from one to five days, according to the *Polar Manual*, depending on the severity of the burn. Dr. Hedblom has found that most victims are more susceptible to another exposure than they were before the initial one. Apparently the conjunctiva remains more sensitive than it was before to ultraviolet radiation. This heightened sensitivity seems to last for a period of from five to seven years.

Snowblinding or sunblinding conditions are general in the Antarctic whenever the sun is above the horizon. It does not matter whether there are clouds. In fact in cloudy weather the incidence of snowblindness reportedly is greater than that when the sun is shining. Men forget about their eyes in cloudy weather, but the ultraviolet radiation is still strong. In the Antarctic one must wear sunglasses all the time out of doors when the sun is up.

The Navy has been recommending glasses of neutral gray with double-gradient density of "inconel" metallic coating.* This type of sunglasses incorporates the Eskimo slit-goggle principle, which reduces glare from both the sky and the snow. I have used a pair of

* These are the Bausch & Lomb G-15 glasses, which the Navy recommends specifically.

green glasses with double-gradient density in the Antarctic and found that, in addition to satisfactory protection from snowblinding, it also improves light-shadow discrimination, and hence makes a good sunglass for any part of the world.

Frostbite is another common hazard. It can come on suddenly after a surprisingly brief exposure. The onset is not usually noticeable to the victim, but it is visible to an observer as a sudden blanching of the skin on the nose, ears, or cheeks. The victim may feel nothing more than a sudden tingling, or "ping" sensation.

First-degree frostbite is similar to sunburn. Tissues become red when thawed and later peel. In second-degree frostbite, blisters form twenty-four to thirty-six hours after thawing. The *Polar Manual* maintains that quick thawing of any freezing injury in a water bath of 100 to 118 degrees Fahrenheit is the best way to speed recovery. But, it warns, this should not be attempted without medical supervision.

Cold feet

The treatments recommended by Dr. Hedblom for frostbite are applicable in any cold climate.

In the field, treat frostbite of the face by placing a warm hand over the spot until it starts hurting, which means it has thawed. A frostbitten hand can be warmed under the opposite armpit. Parkas should always be roomy enough for this. If they are not, they are too tight. Frostbitten feet, according to Dr. Hedblom, "are best thawed on the warm belly under the parka of a trail mate."

Never treat frostbite by rubbing the affected place with snow or slush, Doc warns.

Freezing injuries are another problem. The *Polar Manual* recommends that if a hand or a foot is frozen in the field, rapid rewarming should not be attempted without medical help. According to the *Manual,* it is better to keep the extremity frozen an additional four to eight hours if there is a chance of reaching a hospital or first-aid station in that time than to attempt anti-freezing therapy without supervision and proper equipment.

The man with a frozen foot is not a stretcher case. He can hobble along on it. Said Dr. Hedblom: "Men with frozen legs have walked miles and days for help. But a man with a thawed foot—that man *is* a stretcher case."

While the patient with a frozen extremity is being brought to an aid station, his core (inner) temperature should be warmed with hot fluids, such as soup, chocolate, or coffee, but no alcohol should ever be given in these cases.

Illusions on ice

In the Antarctic men are subject to a number of optical illusions, including mirages, which can be threatening. Most of them, however, are merely interesting and are entirely objective, so that they can be photographed.

Antarctic mirages are quite common. They are caused by a temperature inversion—a warm stream of air layered above cooler air. The inversion reflects light like a mirror and may show images of the landscape over the horizon from the viewer. At the Naval Air Facility on Ross Island it is common to see the peaks of the Trans-Antarctic Mountains across McMurdo Sound displaced from their bases and shifted upward or to the north or south. A cloud bank may obscure the top of one range and just above it the peaks of another range may appear, slightly off center, of course, but with startling clarity.

The sun shining through ice crystals in the clouds produces "ice" bows of brilliant colors, and sometimes they cast a greenish tint to the sky, giving the landscape an unearthly backdrop.

The ice crystals in the atmosphere glow with reflected sunlight, creating the illusion of many little suns in the sky. These "sun dogs" are so brilliant you do not look at them directly.

Of all the optical effects, the most menacing is the whiteout, like the one described earlier. It comes up very quickly and creates the impression that you have suddenly stepped into the nth dimension, where earth and sky have vanished and the universe is a bowl of milk.

Whiteout is explained as the result of the double reflection of light between clouds and snowscape. It causes a diffused, nonpolarized illumination which casts no shadows. Contrast suddenly vanishes, and its importance in visual perception is keenly evident when this happens. Everything becomes pearly gray. There is no horizon. There is nothing.

When you are caught in a whiteout your visual perceptive apparatus becomes unhitched, and you tend to become unhitched from the environment. Some men become dizzy. You cannot judge distances and you cannot see surface detail. A yawning crevasse may stretch directly in front of you, but you may not see it, nor can you see a small hole in which you could stumble and break a leg. What you do in a whiteout is sit down on the ice and wait. After a while the scenery returns, as though someone had raised a window shade, and you are back in the real world once more.

The womanless world

Men who remain in the Antarctic for any length of time usually grow beards. As one might expect, beards are status signs. They help identify the wearers with the environment and its glamorous history —although early explorers, including Scott and Shackleton, are pictured as clean shaven.

On a practical level, a beard helps keep the face warm. But it can be argued that a beard also conceals frostbite. On the trail most men don't want to bother shaving, which uses up time and fuel—both precious commodities.

Dr. Hedblom had this to say on the subject of beards:

One school insists that a clean shave is good for morale. Whose morale? The important thing about beards is they are a matter of individual taste and preference. No one in a position of leadership or command should force his ideas on his associates, one way or the other. This does destroy morale.

As the growing of beards might suggest, Antarctic exploration is a male enterprise. Antarctica is a womanless world. The Navy has no signs up advising women to stay out of its stations on the ice. It simply declines to transport them to the region. Navy officers have shown an understandable reluctance to discuss the subject, and about all they have been saying is that it is against Navy "policy."

Generally, the older Navy officers feel that the presence of women —"lady scientists" or even nurses—would "impair efficiency." One of the military engineers at McMurdo Sound complained, "We'd have to build another head [latrine]." Others oppose women on the ice on such grounds as "We don't have clothing to fit them," or "We'd have to give them separate quarters and we're overcrowded now," or "You'd be amazed if you knew how complicated the logistics of providing for females can be."

Navy enlisted men are dead set against allowing women on the ice, according to the informal "soundings" I have taken. Their reasons range from masculine pride in keeping the ice cap the last untram-

A Navy cook inspects bread in the McMurdo Station galley.

A Jamesway hut serves as the chow hall for personnel at Williams Field.

meled sanctuary of the male, to a suspicion that if women were added to the station complement "the officers would get them." The enlisted men showed considerable apprehension that they would be rejected by women on account of status, and one way to keep that issue from arising was to keep women out of Antarctica.

One high-ranking Navy officer at McMurdo Sound said he would approve female personnel only if there was one for every man.

The scientific community on the ice appears to be somewhat divided on this question. A few scientists whose wives are also trained researchers would like to bring them, but are conscious of the housing and morale problem this would create.

In contrast to the Navy's position, the National Science Foundation's Antarctic Programs Office makes provision for women scientists aboard its oceanographic research vessel *Eltanin*. Quarters aboard the vessel are considerably more crowded than at an Antarctic station, but no one has complained that the presence of women has complicated shipboard living unduly. The *Eltanin* sails the Antarctic seas but does not approach land in the region.

Rear Admiral George J. Dufek, now retired, who commanded the Antarctic Support Force during the period of station construction before the IGY and during the IGY, had this to say: "I was against women in the Antarctic for the simple reason that during the construction period there was no place for them. So far, this has been a man's world, with two or three exceptions."[4]

One exception was a visit by two airline stewardesses on the only commercial airliner to land in the Antarctic—a Pan American World Airways transport. The airplane flew into McMurdo Sound from New Zealand on October 15, 1957. The girls, Patricia Hepinstall and Ruth Kelly, were wined and dined in the Naval Air Facility mess hall. They even judged a beard contest.

Admiral Dufek noticed a good many of the men were missing, and when he looked around the camp he saw they were sitting in their quarters.

"None had gone to meet the plane," he related. "They had remained indoors when the girls drove up Main Street. They wanted to be able to say that from the time they left civilization until they returned they hadn't seen a woman."

By 1965 the only women who had wintered over in Antarctic regions were Mrs. Edith Ronne and Mrs. Jennie Darlington, who accompanied their husbands on Finn Ronne's 1947–1948 expedition, which was based on Stonington Island off the coast of the Panhandle.

Dufek predicted that "eventually" women will come to Antarctica. Until they do, it is not likely that the region will be occupied extensively. Nor is it likely, either, that the style of occupation will advance much beyond the outpost stage.

Since the emotional and behavioral problems of long-termers are essentially the products of this level of habitation, it seems reasonable to speculate that allowing women at Antarctic stations would tend to reduce the effects of isolation, and the problems that go with it.

White nights and black days

The environment itself is not without its effects. It contributes to several psychosomatic complaints, which seem to be related to the difference in light. For about fourteen weeks of the year at McMurdo and Byrd Stations and for nearly twenty-four weeks at the pole the sun is either continuously above the horizon or below it.

The periods of total light and total darkness vary with the distance from the geographic south pole. Curiously, continuous sunlight and

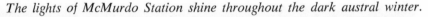

The lights of McMurdo Station shine throughout the dark austral winter.

continuous darkness both inflict severe insomnia on some men. When the sun is shining twenty-four hours a day, these men complain that they cannot sleep for that reason. In the winter they believe they can't sleep because it is dark all the time.

During the winter dark periods, some of the insomniacs will awaken after an hour's sleep and be unable to go back to sleep again. They will shuffle through the station, looking for something to read or for some work to do in the shops. Psychologists have related the insomnia to anxiety, but since the victims have no access to treatment they simply have to live with it, which they do.

On the trail, oversnow traverse parties have reported that morale sinks when the going becomes difficult because of storms or crevasses. It bounces back up again, however, after the crevassed region has been negotiated and the weather clears.

The outpost society

The overriding concern in psychological adaptation to Antarctic living is interpersonal relationships. These are affected by the size of the station, especially for wintering-over parties.

Generally, as we shall see in a moment, the smaller the station the more difficult interpersonal problems are likely to be. I would expect to see similar problems arise in the exploration of space, where, again, small groups of men will venture forth away from the mainstream of society and will have to adjust to isolation.

One of the lessons the Antarctic experience has taught us is the necessity for screenout of potential troublemakers—men whose personality difficulties do not let them adjust readily to outpost society. This became evident during Operation Deep Freeze I (1955–1956). Deep Freeze was the code name the Navy used for its support of science on the ice—and still uses. After Deep Freeze I, when one man developed a psychosis and had to be evacuated, military-service volunteers for Antarctic duty were screened by a battery of psychological tests and by interviews with both psychologists and psychiatrists.

Of seven hundred Navy men assigned to the Antarctic in Operation Deep Freeze II (1956–1957) and Deep Freeze III (1957–1958), only six were evacuated for psychiatric reasons.[5] None developed a psychosis, but all showed symptoms of disabling emotional disturbances within one to four weeks after they arrived on the ice.

It was found that the size of the population and the length of a man's stay at a station affect his adjustment to isolation. At small

Above: *Base chapel at Williams Field, built during the winter season between Operation Deep Freeze I and II.*

Below, left: *A Navy chaplain with his "adopted" penguin chick which had lost its parents at the Cape Crozier rookery.* Below, right: *A dentist working on a patient during the long winter day, and* (bottom) *galley food supplies in an underground tunnel at New Byrd Station.*

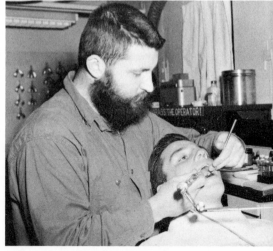

Above: *Many buildings are adorned with amusing signs such as the one on the officers' quarters at McMurdo Station.* Below: *Burma Shave signs add a playful touch of color in the white landscape at Williams Field.*

stations, such as Amundsen-Scott or Byrd, men of the wintering-over party tend to develop a family-like set of relationships, the psychological studies have shown. Strong feelings of interdependence develop because each man is needed to do an important job. These feelings inhibit friction and may account for the fact that in sixty years of Antarctic exploration there is no record of murder or aggravated assault. In a station as large as McMurdo (actually, it is the largest in the Antarctic at this writing), where the summer population may exceed 1000 men and the winter crew is 135, there is a tendency among the men to form cliques, so that a man may have a choice of groups to belong to.

The main problem in the small station is managing to get along with just a few other people representing a limited number of personality types over a period of time ranging up to nine months. In the big station, however, a man is not dependent on a single group for his social needs.

The concerns of men staying only for the spring and summer are different from those of the men who winter-over. Short-term Navy personnel worry about the absence of liberty ports. The long-termers repress these feelings of deprivation and are more concerned with the quality of food and type of job they are expected to do.

New arrivals among Navy personnel show anxiety about what is expected of them. The anxiety is relieved by a work assignment, which gives the man his "anchor." Work is the most meaningful of all the roles in the support force, and adjustment studies show that the more fully a man is able to occupy himself working, the better he is able to adjust to isolation.[6]

Some men take on more than one job, as a means of keeping occupied and of increasing prestige. At Pole Station, the Navy commander may be a physician who also undertakes the chore of housing officer. At McMurdo Sound, the dentist may be the financial officer and the chaplain may operate the post exchange.

When winter comes and the sun goes down, Antarctic stations become closed societies. Traditional roles tend to lose their importance and give way to the roles which are relevant to outpost living. So far as survival is concerned in the station, what status a man has in the outer world becomes meaningless. Strong personalities in the group of Navy personnel and civilian scientists come to the fore.

At one station, conflict between the nominal leader and the natural leader among Navy enlisted men was noted as a cause of serious dissension and disruption in the group. In this situation, the apparent source of trouble was the reluctance of the scientists to share house-keeping chores with Navy men. The scientists contended that their physiographic, meteorological, upper-atmosphere, and aurora-research activities kept them occupied full-time. It had been their understand-ing that the sole task of the Navy men was to maintain the station. The Navy unofficial leader contended there was too much work for military personnel. This situation never was resolved, according to narratives I heard from men who were involved in it.

There have been a number of "incidents" at stations involving verbal and some minor physical conflict, but none has led to charges against anyone. I have interviewed the participants in what the social psychologists might call one "conflict situation." The two men threw a few punches and a large volume of verbal abuse at each other. At the time I talked to them, they were preparing to leave Antarctica and there were then no signs of hostility between them. Neither could give me a clear picture of what the trouble had been between them or how it had started. They kept saying, "I guess we just got on each other's nerves."

I have talked to scientists who behaved in a similar way during a wintering-over period. Their conflict was more subtle, characterized by exhibitions of "studied indifference" to each other. One said he would "think up questions" in the field of his adversary in order to "trap him and show him up." He said the other man was pompous and egotistical, and needed "deflating."

Dr. John H. Rohrer, Georgetown University Medical School psy-chologist, has reported that the VIPs in a small station tend to be the cook and the radio operator. He explains:

The cook who is directly related to one of the few socially approved gratifications permitted the men achieves far higher social placement within the group than do the scientists who are making observations. The radio operator who maintains contact with the world outside the camp is held in similar esteem.[6]

During the winter, the ionosphere permitting, men at McMurdo Sound, the pole, and other stations can talk to their families back home by radio telephone. At NAF McMurdo, Navy radio men operate the KC-4-USV radio station, the largest broadcasting unit that the United States has in Antarctica, as a ham radio at certain hours for direct personal communication service.

Station KC-4-USV transmits and receives on the 20-meter amateur band. The operators have arranged weekly broadcast schedules with "hams" in the United States who can "patch" the signals from Ant-

An exterior view of the radio ham shack at McMurdo Station.

An amateur radio operator using the microphone at Station KC-4-USV.

arctica into a telephone circuit. The result is a fairly good two-way conversation which costs the Antarctican only the price of a collect telephone call to his home from the amateur radio operator's telephone.

Psychological adjustment to the Antarctic proceeds in several stages, which observers have attempted to define. The first phase covers a brief period during which the new arrival becomes accustomed to the landscape, loses his fear of the cold, and accepts the continuous sunshine.

A second phase of adjustment appears when the sun sets for the winter, it becomes "real cold," and the wind moans and shrieks between the huts. Then the first signs of depression show. Men begin to wonder whatever brought them to this forlorn place. At this point, symptoms of uneasiness appear. More of the men are sleepless—and for longer intervals. Some, according to Rohrer, get no more than one hour's sleep a day for seven or eight days at a stretch. Some develop symptoms of sensory deprivation, such as hallucinations and inability to concentrate or pay attention.

Everyone seems to become more irritable with the onset of winter. The irritability, however, is fairly well repressed. Some of the men complain of frequent headaches, usually more severe and longer-lasting than any they have experienced before. "Antarctic headache" has been analyzed by a number of medical men and psychologists. Station doctors say that generally they cannot find a physiological basis for it and tend to view it as psychosomatic. Observers have concluded that the headaches occur more frequently among the "intellectuals" than among "non-intellectuals" at the station—"intellectual" being defined as a man with a university degree.

Antarctic headache is a thinking man's disease, the studies indicate. It may be a manifestation of repressed hostility. Young, comparatively uninhibited men can work off hostilities and aggression in horseplay, but the older, more restrained "university types," both Navy and scientist, do not use this outlet. It does appear that the youngsters have fewer headaches than the intellectualizers.

In winter regressive behavior appears in some men. They seem to act as they may have done at a younger age. This kind of behavior tends to become more pronounced after Midwinter Day, June 21, actually the start of winter by the calendar.

One form of extreme regressive behavior is sleeping eighteen hours a day. Some men listen for hours to phonograph records without moving, some become insensitive to sounds, and others can't bear certain noises. Some of the men lose all interest. They spurn the hobby shops and books. They prefer long bull sessions, cards, or some other passive kind of amusement.

Lecture courses are probably the most profitable and popular activities during winter recreation hours. Scott inaugurated this tradition at Hut Point during the 1901–1904 *Discovery* expedition. It has been carried on by Antarcticans ever since.

Navy enlisted men who attended the courses at Byrd and Pole Stations have said they gained new insight into the entire Antarctic experience through the lectures. They also learned something about mathematics, meteorology, ionospheric physics, and marine biology. The "professors"—scientists who volunteer to give the lectures— spend a good deal of time preparing them.

Midwinter Day is observed by all Antarctic stations, and has been a special day since Scott's time. The President of the United States usually sends a message saluting the men on the ice, and the day frequently winds up with a roaring good party.

In the year 1959 the monthly summary report radioed to Washington from United States stations in the Antarctic revealed that the men held a Midwinter Day observance at Byrd Station, where, the report added, snowdrifts were so high that extra lengths had to be added to the kitchen stovepipe. The snow was a foot above the roofs of the huts. Since 1961, of course, the station has been dug in under the ice so that no obstructions protrude above the snow surface to catch blowing snow and raise drifts.

"A remarkable Midwinter's feast and celebration" was reported in 1959 at Amundsen-Scott Station, although the message did not specify what was remarkable about it. All seventeen men wintering over in the station that year took part. The report added that water was getting to be a problem. In these "status reports" from the stations, joys and tribulations are intertwined for the sake of brevity.

"Twice a week," the Pole Station report said, "over five tons of snow must be dug by hand for the snow melter in temperatures of 90 degrees below zero and winds of 20 miles an hour—all at an altitude of 10,000 feet above sea level."

The snow is mined for water, into which it is converted by the snow-melting device behind the kitchen. In order to avoid contamination (by hydrocarbon wastes from the Diesel engines), the snow is dug at some distance from the camp in a minelike shaft under the surface. Even with this precaution, all kinds of debris turns up in the snow-melter sieve, ranging from volcanic ash at McMurdo Station to teardrop-shaped blobs of meteoric iron at Amundsen-Scott. The blobs apparently were bits of iron meteorites which broke apart and melted, during entry into the atmosphere, and then congealed on hitting the ice.

One of the classic "gags" in modern Antarctic lore was a carefully contrived plot to convince reluctant members of the water detail that there was gold in the snow mine at McMurdo. One of the cooks was seized with the idea when he saw some shiny scraps of copper glinting on the machine-shop floor. He picked them up, melted them, and then cooled them into shiny blobs no bigger than a beebee. Taking others of the kitchen staff into his confidence, he managed to spread the word that some shiny yellow metal had been found in the snow-melter sieve, and that one of the "docs" (scientists) had identified the stuff as gold, doubtless blown out of Mount Erebus, the smoking volcano on Ross Island, during some past eruption.

The snow-mining detail paid little attention to this rumor until men cleaning out the snow melter found the "gold" pellets in the sieve. There was no shortage of volunteers for the snow-mining detail for several days, until the hoax was uncovered.

A psychological study made of men who stayed a year at one station has revealed that general hostility toward the leader may be a positive force in keeping a group functioning.[7] The study involved a wintering-over party of thirty-nine men—five officers, twenty-five enlisted men, and nine civilian scientists. No psychoses or disabling depressive states developed. Physical health was good—so good, in fact, that absence from duty for medical reasons was rare.

Men showed personality disorders, but these did not interfere with the ability of anyone to carry out assignments. One schizoid man lived and slept away from the others. He rarely talked with anyone else. Yet he was regarded as one of the more capable and valuable members of the station.

In the opinion of observers, "one of the dynamics which kept the group functioning was an antithetical attitude between the station leader and the other men." While he was highly able and experienced, the leader was regarded as being "inflexible in his interpersonal dealings." The report added:

He provided a convenient and plausible focus for the hostilities inevitably generated in a group of men whose interests and backgrounds are disparate and who are thrown together for a long time in close personal associations.

The group managed to achieve cohesiveness because of their general feelings toward the common source of grievance—the leader. But overt hostilities were rare. One abortive fist fight was reported. Violent, angry arguments were practically nonexistent, although the beer supply was liberal. Each man recognized that he was dependent on the good will of the next man in this tight little world for his own feelings of acceptance, security, and worth.

The study found that Navy enlisted men felt disillusion when they found that station life was hardly the democratic Utopia they had expected at such an outpost. They were still under discipline and there were plenty of chores. Civilian scientists were disenchanted, too. They had expected an academic atmosphere, free of military interference and chores that would distract them from their research. Instead, pressure was put on them to cooperate in work details.

After several months on the ice about one-third of the men began to show distinct lapses of memory. They had difficulty concentrating. As the winter wore on the distractibility and absentmindedness increased. Mild "fugue" states appeared in which mental organization seemed to disintegrate. Men said they could recall leaving their quarters and then suddenly would "come to" in another part of the station without being able to remember how they got there. Two of the men grew so preoccupied and inattentive that they became a source of annoyance to others in the station. "This impairment of memory, alertness, or awareness was a rather striking phenomenon and requires more study," the Navy research men said.

Isolation from women was not in itself a problem for the group or for any individual, the report said.

Toward the end of the wintering-over cycle two cliques appeared among the technicians—four electronics men versus four aviation mechanics. The techs refused to speak to the mechs, except in line of duty. No one was able to relate afterward why the split began. Some of the men said it continued under inertia. Men who believed it was silly hesitated to say so for fear of being ostracized by their own cliques.

The fugue states and distractibility were only temporary and cleared up when the men were relieved from duty to go home.

Stress of isolation was not troublesome in itself, the investigators reported. There was frequent contact with other stations and with families at home by the short-wave radio, which is regarded as an important morale booster.

During the winter of 1960 an international chess tournament was organized on the ice cap. Americans at McMurdo Sound and New

Zealanders at nearby Scott Base played the Russians at Mirnyi and Lazarev Stations, two thousand miles away on the great eastern ice sheet. Sven Evteev, the exchange Russian scientist at McMurdo, served as translator for McMurdo and Scott Base players.

I have been unable to find any record of who won. In the austral spring of 1960 I encountered Professor Evteev at McMurdo. He was working on a Sno-Cat in a big shed, where the vehicles for Albert P. Crary's traverse to the south pole were being assembled. I asked Evteev who had won the south polar chess sweepstakes.

"Who knows?" he responded, diplomatically. That was all I could get out of him.

The third phase of adjustment to Antarctica appears after Midwinter Day. The men begin receiving definite signals that the period of isolation is ending. McMurdo Sound reported:

> The week following the Midwinter Day was quiet, yet filled with the normal industry of men completing inside tasks and routine camp details. The growing halo of light to the north and east heralds the rising sun, and the solitude of winter has begun to break.

Now as they venture out of the huts in the cold dark, men look north, where the red glow of the returning sun becomes more brilliant every day.

At last the sun rises. Aircraft appear. The first C-130 Hercules aircraft, successors to the C-124 Globemasters that pioneered the region, lumber into McMurdo Sound from New Zealand. They receive a wild welcome. They bring fresh fruit, fresh mail from home, and fresh liquor. They are harbingers of spring.

Winter is officially over. The sun is up, blinding in the sky. Suddenly there is beauty in the frozen world. The men begin to lose that feeling of being adrift in space. In the sunlight the magnificent aurora fades out. Depressions and regressive behavior disappear or at least become less pronounced. However, headaches may increase in frequency at this time. According to Rohrer, the men say the headaches are due to sunlight. But he believes they are probably a result of "unverbalized wishes to get off the ice."

The final phase of adjustment appears before the men are due to board ship for New Zealand and home. Work patterns fall apart. The men seem unable to complete tasks, especially where they are required to repair or maintain equipment. They have unconsciously separated themselves from dependence on the station machinery, and maintenance becomes less meaningful.

Once they sail north across the Antarctic convergence into normal day and night, the headaches disappear and the insomnia is gone. So are most of the hostilities generated on the ice.

According to three investigators,[8] the psychological screening has been successful in keeping troubled men out of Antarctica. From 1956 to 1962 there were no documented cases of psychiatric illness reaching psychotic proportions. Also during that time not a single case of psychosexual disturbance or overt homosexual activity has been reported among the Antarcticans. "Functional" backaches and gastrointestinal complaints, which make up the bulk of sick call at military posts in most places, were rare in Antarctica.

This does not mean that acute emotional disturbances did not appear. When they did, others in the group tried to help the afflicted man and tended to treat him as an ill person. Their consideration may have pulled through the winter many a man who might not have made it otherwise.

Of the six cases of psychiatric illness mentioned earlier, five were last-minute replacements who had not been screened. These men suffered from severe neurotic depression and anxiety reaction, the psychologists reported.

In the opinion of these psychologists, the only neurotic mechanism which should disqualify a man from Antarctic duty is extreme rigidity. "The rigid person who maintains ego integrity by adherence to fixed beliefs or routines, without flexibility in the face of group needs, invariably becomes a source of disruption to the group during wintering over," they said.

Demanding, sensitive, narcissistic individuals usually became a source of friction and a demoralizing influence, the report noted. Some of these men could get by in a larger station but caused trouble when they were thrown with a small group.

Antarcticans versus astronauts

A psychologist whose interest in Antarctica led him to accompany the Van der Hoeven Traverse in Victoria Land in 1959–1960 has compared Antarcticans with astronauts and finds them similar in self-motivation. He is William M. Smith of George Washington University. Working with Marshall B. Jones of the United States Naval School of Medicine, Pensacola, Florida, Smith tested astronauts, Antarctic scientists, Navy cadets, and Navy retrainees on the Pensacola Z scale.

The Z scale is a sixty-six-item questionnaire which reveals a man's degree of autonomy or self-direction. Each item consists of two questions describing personality traits, and the subject is asked to choose which one more nearly fits him. One item, for instance, asks the subject whether he is (a) anxious or (b) conceited. The answer indicating autonomy or self-reliance is "conceited." Another asks the subject whether he is (a) dogmatic or (b) sloppy. The response that he is "sloppy" indicates that the subject tends to be nonrigid and flexible.

The astronauts consisted of 26 of the 31 men who were given psychological tests for Project Mercury. The scientists were 57 civilians who had wintered over in Antarctica in 1957–1958. The cadets were 766 college sophomores who had reported for naval air training at Pensacola, and the retrainees were prisoners—407 of them—in the "brig" at the Navy Retraining Command, Portsmouth, New Hampshire.

Astronauts and Antarctic scientists scored high in personal autonomy. They described themselves in "far more autonomous terms" than did either the cadets or retrainees. The retrainees scored the lowest in terms of self-motivation. Some of the difference could have been accounted for by intelligence. The cadets, for example, displayed an average intelligence quotient of 110, while the astronauts' average on the Wechsler Adult Intelligence Scale was 133.

The experiment showed what one would expect. Shackleton said it sixty years ago: "Men whose desires lead them to the untrodden paths of the world have generally marked individuality."

A land without passports

Now the untrodden paths are being trod by an ever-increasing population of investigators. Under the Antarctic Treaty of 1959 no one requires a passport of any kind to enter Antarctica. The treaty is the first in the annals of mankind to set aside a continent for science.

In this way, the ancient concept of Antarctica has become a guiding influence in diplomacy, as well as in science. The Antarctic experience has shown men a profound mystery of the earth. It has shown men the mystery of themselves. In an intensely practical way it has demonstrated the advantage of international cooperation in the investigation of new frontiers.

The Antarctic Treaty provides that the continent will be used for peaceful purposes only. It permits overflight and inspections of all bases. It bans nuclear explosions and military preparations on the ice cap. Duration of the treaty is indefinite, but after thirty years any of the twelve signatories may call a conference for review and amendment. The treaty was signed December 1, 1959 by Argentina, Australia, Belgium, Chile, France, Japan, New Zealand, Norway, the Union of South Africa, the Soviet Union, the United Kingdom, and the United States.

There is a growing belief among American scientists and politicians that the treaty will be the prototype for agreements covering the exploration of the moon. However, the Antarctic Treaty followed sixty years of exploration on the southern ice cap. By that time the nations were convinced that whatever mineral wealth lay buried in the shield of East Antarctica was well guarded by two miles of ice. Antarctica was dedicated to science only after it became obvious that the region was of little use for anything else. It is distinctly possible that advancing technology may change that view.

In the early years of the century Australia, France, New Zealand, Norway, and the United Kingdom had staked claims on pie-shaped segments of the continent. NAF McMurdo on Ross Island lies within the Ross Sea Dependency of New Zealand, a region about the size of midwestern United States. Economically, the region appears to be as useful to the thrifty New Zealanders as the Land of Oz. Yet this could change. There has been a proposal to establish an airport at Marble Point on the coast of Victoria Land, just across McMurdo Sound from the Naval Air Facility. It would serve as a refueling stop for transpolar commercial and military flights. It would probably stimulate air traffic in the rapidly developing commerce of the southern hemisphere. The cost has been estimated at $200,000,000—and that is as far as the project has gone.

However, atomic power was introduced into Antarctica in 1961 with the installation at McMurdo of a nuclear reactor, fueled with enriched uranium. The atomic-power plant generates 1500 electrical kilowatts. By 1964, it had produced 6,400,000 kilowatt-hours of electricity during tests.

New Zealand uses the dependency as the site for Scott Base, the efficient station near the American base. It also employs the scenic attractions of Mount Erebus, the smoking volcano, and the local penguins on a series of handsome postage stamps.

Argentina and Chile came late to the refrigerator continent and made claims which conflict with each other and with those of Great Britain.

Altogether, the claims by the seven countries cover about four-fifths of the ice cap, leaving one-fifth, Marie Byrd Land, for anyone who wants it. The United States has explored more extensively in that region than any other country, but never put forward a claim. That is just as well. As Bentley and Anderson showed, Byrd Land is mostly ice resting on sea bottom.

The Soviet Union has never put forward a claim in the Antarctic either. The Russian government stated in 1950, however, that it would not recognize any Antarctic settlement unless it was a participant. Later, in 1958, the Soviets stated they retained all rights based on discoveries by Soviet citizens—including the right to exert a claim. Japan renounced Antarctic claims in 1951.

The big diplomatic headache in the Antarctic is caused by the overlapping claims of Chile, Argentina, and Great Britain to territory in the Falkland Islands, the South Shetland Islands, and the Panhandle.

The South American countries base their claims on geographical proximity and on inheritance from Spain of a fifteenth-century title to undiscovered lands in the western hemisphere. This derives from the division of new lands between Spain and Portugal by Pope Alexander VI in 1493.

The Pope drew an imaginary line of demarcation 100 Roman miles (leagues) west and south of the Azores and running from pole to pole. He assigned all lands lying west of the line to Spain and all newly discovered territory to the east of it to Portugal. Later the line of demarcation was moved by the two powers to a point 370 leagues west of the Cape Verde Islands. The arrangement averted a naval war and set the pattern of colonization by the two countries for 300 years, with Spain dominant in all of South America except Brazil.

Chile additionally bases its claim to the Panhandle on geophysical evidence that it is a continuation of the Andean Cordillera. In 1947 Britain proposed that the International Court of Justice settle title to the overlapping claims, but Argentina and Chile declined to accept the Court's jurisdiction in the dispute. Britain tried again to take the matter to the Court in 1955, but the other parties would not agree.

On March 26, 1956, it was stated in the House of Commons that the United Kingdom government "now resume their full freedom to take whatever further action may be required to maintain their title."[9] Meanwhile the British offer to submit the question to the International Court remained open.

During the IGY territorial infringements were overlooked if the trespassers were connected with the scientific endeavor. Britain sent notes concerning the landing of Soviet scientists and sailors in the

South Sandwich Islands in December 1957, and the visit of an Argentine vessel with a party of tourists to Deception Island in the South Shetlands in January 1958.

Some kind of general settlement was in the air for Antarctica immediately after World War II. No one had forgotten a foray by Hitler's navy into the region in 1939, presumably for the purpose of scouting a submarine base. In 1948 the United States proposed an eight-power condominium for the continent, but only Britain and New Zealand showed any interest.

Ten years later, with the IGY in full swing, the future of Antarctica was taken up by Prime Minister Harold Macmillan with the prime ministers of Australia and New Zealand. They discussed, Mr. Macmillan said, "means of ensuring that Antarctica did not remain a potential source of friction and conflict." They had agreed, he added, on certain basic principles, such as the free development of science in Antarctica and the necessity of insuring that the continent would not be used for military purposes by any nation.

Argentina and Chile stood pat on their territorial claims. But they showed interest in scientific cooperation.

When the IGY ended, it was evident that the surface had merely been scratched in Antarctica so far as science was concerned. The United States then proposed a treaty guaranteeing continued investigation in the region and raising the trial balloon of a nuclear-test ban and overflight inspection.

The twelve nations with interests—territorial, scientific, or both—agreed after hardly more than sixteen months of discussion. The treaty was signed at Washington, D. C., on December 1, 1959. Its preamble contains a statement unique in the history of nations: "It is in the interest of mankind that Antarctica shall continue forever to be used exclusively for peaceful purposes and shall not become the scene or object of international discord."

In the United States Senate ratification of the instrument came up for debate on August 10, 1960. It was a sultry day in Washington, humid after an early-morning rain. The late Senator Clair Engle of California moved that further consideration of the treaty be postponed until January 25, 1961. He scarcely needed to remind his colleagues of the November election. He said:

The new administration which will have the responsibility of conducting the foreign policy of this nation for four years . . . should have the opportunity of passing on the provisions of this important treaty. The treaty involves an area as big as the United States of America plus all of Europe.

For all practical purposes, it disposes in perpetuity of the relationship of this nation and other major nations to the vast continent of Antarctica.

Engle raised a number of doubts about the treaty and called for more time to resolve them. What reason was there, he demanded, to believe that the Russians would abide by the treaty terms?

These remarks were the signal for Southern Democrats to attack the treaty as a "giveaway" of American rights in Antarctica. Liberal Democrats from the North and West and Eisenhower Republicans rallied to the treaty's defense. Early in the debate, it became evident that this coalition held a majority, but whether it could muster the two-thirds vote required for ratification of a treaty was in doubt.

Old fears and suspicions clashed with the new, practical idealism embodied in the treaty. Southern Democrats were not to be moved by assertions that Antarctica was economically and militarily valueless. Marie Byrd Land, where the United States could make a claim, was worthless, it was argued. The region was just a mile-thick cake of ice.

How does anyone know it will remain worthless? demanded Senator Richard B. Russell of Georgia. "In this atomic age when someone may develop a means, such as by atomic power, to burn off some of the ice cap, one could not say that area is completely worthless."

Senator J. William Fulbright of Arkansas, chairman of the Senate Foreign Relations Committee, addressed himself to that point by quoting a report by the Washington Senator Henry M. Jackson, who had recently visited Antarctica, that if the ice cap melted it would raise the level of the Pacific Ocean by a hundred feet. "It would submerge most of the State of Washington," Fulbright observed in his drawly Arkansas style, "and he also said he did not think he would be in favor of that." Moreover, he added, scientists who testified before the Foreign Relations Committee had characterized such a proposal as "wholly unrealistic."

Engle argued that the Russians would be the only beneficiaries of the treaty. He feared they would move in and take over. "When the Soviets went into Antarctica during the IGY," he said, "they squatted on an area which had been claimed for many, many years by Australia and they established their camps there."

"That brings up another aspect of the problem," said Fulbright. "As a matter of fact, the land where we have squatted belongs to New Zealand, if we recognize traditional claims."

Engle's motion to postpone action until January was defeated, 56 to 29.

In defense of the treaty, Senator Frank Carlson of Kansas read an analysis of it prepared by Herman Phleger, the State Department expert who had led the United States delegation in negotiations. Though at the time it went virtually unnoticed, the analysis is one of the unique statements in modern diplomacy:

This treaty for the first time in history devotes a large area of the world to peaceful purposes. It is the first treaty which prohibits nuclear explosions with adequate inspection.

It is the first treaty to provide freedom of scientific investigation over large areas and it constitutes a precedent in the field of disarmament, the prohibition of nuclear explosions and the law of space.

On the final vote the treaty was ratified, 66 to 21.

From concept to conquest, it has taken two and one-half millennia to realize the ancient dream. Antichthon, Antarktikos, Terra Australis, and Antarctica are symbols of intellectual evolution, each signifying a stage of environmental awareness and technological capability, and successively revealing a pattern in the expansion of mankind.

Genetic evidence, the biological scientists tell us, suggests that the present human species originated in one place and spread over the earth. Only in our time has the long migration reached the last land frontier on the planet.

Man now inhabits the seventh continent of the world. The lights of science stations that shine on the ice in the dark of the south polar winter from the Eights Coast of Mac-Robertson Land, from Cape Hallett to the pole, are not likely to be extinguished, since many years of research remain to be done in the ice-age laboratory.

Even so, a new cycle of exploration has begun. As the ancients conceived of regions beyond their inhabited world, which they called the *Oikoumene,* so do we conceive of new lands beyond our world, in space. The thrust of intellect is the same and only the definition of the *Oikoumene* has changed: in the modern age the environment of man encompasses the solar system.

Antarctica has been the first of the great environmental investigations made possible by present technology. Space is the next. A new adventure looms across the seas of plasma that surge between the worlds. New lands beckon, not yet known. So it will always be.

REFERENCE NOTES

CHAPTER 1

1. New Zealand Antarctic Society, *Antarctica Today*, 1951.
2. *Edinburgh Philosophical Journal*, April 1821.
3. *Transactions of the American Philosophical Society*, Vol. 31, Part I, January 1939.
4. Captain Edmund Fanning, *Voyages Around the World*, 1831.
5. *Transactions of the American Philosophical Society*, Vol. 31, Part I, January 1939.
6. L. P. Kirwan, *A History of Polar Exploration*, 1962.
7. *Proceedings of the Royal Society of London*, Vol. 62, No. 387.

CHAPTER 2

1. U.S. Coast Guard magnetic pole survey, 1960.
2. Kirwan, *op. cit.*, p. 289.
3. John Lilly, *Man and Dolphin*, 1960.
4. Interview at Skyland, Virginia, with Sir Charles S. Wright, September 1960.
5. Leonard Huxley, *Scott's Last Expedition*, 1913, Vol. II, p. 286.
6. *Glaciology Report, Terra Nova* expedition, 1922.
7. Lecture by Scott at Cape Evans, 1911.
8. A. P. Coleman, *Ice Ages*, 1926, p. 44.
9. R. F. Flint, *Glacial Geology of the Pleistocene Epoch*, 1945, p. 49.
10. *Geographic Journal*, Vol. 43, pp. 605–630.
11. W. L. G. Joerg, *Geographical Review*, Vol. 26, pp. 454–462.
12. *Geological Society of America Bulletin*, Vol. 46, pp. 1367–94.
13. Office of Naval Research, Ronne Antarctic Expedition, *Geomorphology of Marguerite Bay, Palmer Peninsula*, 1953.
14. In *National Geographic*, October 1959, p. 528.

CHAPTER 3

1. *Report of the Committee on Interstate and Foreign Commerce*, U.S. House of Representatives, February 1958.
2. Vivian Fuchs, *Antarctic Adventure*, 1961.

CHAPTER 4

1. American Geographic Society, *Problems of Polar Research*, 1928.
2. Richard E. Byrd, *Little America*, 1930.

3. Smithsonian Institution Board of Regents, *Annual Report,* 1928.
4. Byrd, *Discovery,* 1935.
5. Frank Press and Gilbert Dewart, "Extent of the Antarctic Continent," *Science,* January 19, 1959.
6. "Structure of West Antarctica," *Science,* January 15, 1960.

CHAPTER 5

1. A. L. Wegener, *Origin of the Continents and of the Oceans,* 1923.
2. Eduard Suess, *Face of the Earth,* Vol. I, page 596.
3. *Ibid.*
4. New Zealand Antarctic Society, *Antarctica Today.*
5. Martin Holdgate, *New Scientist* (No. 239), June 15, 1961.
6. *Natural History Report, Terra Nova* expedition, 1914, comments of A. C. Seward, professor of botany, University of Cambridge.
7. *Scientific American,* September 1962.
8. E. Olson, "Continental Drift," *Chicago Museum of Natural History Bulletin,* September 1962.
9. U.S. Coast and Geodetic Survey, 1955.
10. Smithsonian Institution Reports, *Paleo-Geographic Relations of Antarctica,* 1911–1912.
11. Francis Darwin, *Life and Letters of Charles Darwin,* 1887, Vol. III, p. 248.
12. A. Du Toit, *Our Wandering Continents,* 1937.
13. Griffith Taylor, *Environment, Race and Migration,* 1937.
14. New York University Symposium for science writers, March 11, 1961.
15. J. Tuzo Wilson, *Scientific American,* April 1963.
16. From a lecture by Robert S. Dietz at Northwestern University, Evanston, Ill., October 25, 1962.
17. A. P. Coleman, *op. cit.*
18. *Ibid.*
19. Niagara Folio, U.S. Geological Survey.

CHAPTER 6

1. *Report of the Committee on Appropriations,* May 1, 1957.
2. Kirwan, *op. cit.,* p. 231.

CHAPTER 7

1. Erling Dorf, Museum of Paleontology, University of Michigan.
2. Morton J. Rubin, U.S. Weather Bureau, 1956.
3. W. B. Moreland, *Weatherwise,* December 1958.
4. Rastorguev, and J. S. Alvarez, IGY World Data Center Report Series, 1958.
5. Moreland, *op. cit.*

6. *Hearings, U.S. House of Representatives, Committee on Appropriations,* 1959.

7. H. Wexler, Royal Meteorological Society, *Quarterly Journal,* Vol. 85, No. 365, July 1959.

8. Lecture by Dr. Hedblom, Antarctic Symposium, Skyland, Va., 1961.

9. G. Frederick Wright, *Ice Ages in North America,* 1911.

10. *IGY Bulletin* No. 31, January 1960.

11. Kirby J. Hanson, USWB, *Journal of Geophysical Research,* March 1960.

12. H. W. Ahlmann, *Glacier Variations and Climatic Fluctuations,* 1953.

13. *Chicago Museum of Natural History Bulletin,* Vol. 35, No. 3.

14. Ahlmann, *op. cit.*

15. Lecture, September 3, 1959, York Meeting, British Association for the Advancement of Science.

16. Report by Kirby Hanson, Symposium on Antarctic Meteorology, Melbourne, Australia, 1959.

17. Morton J. Rubin, *Scientific American,* September 1962.

18. Huxley, *Scott's Last Expedition,* Vol. II, *Meteorological Report.*

19. *Report of National Center for Atmospheric Research,* Boulder, Colorado.

20. B. W. Harlin, *Weatherwise,* August 1958.

CHAPTER 8

1. L. Reiffel and C. A. Stone, *IIT Research Bulletin,* Summer 1963.

2. *Natural History Report, Terra Nova* expedition, 1922.

3. H. M. Morozumi, paper presented to the Western Meeting of the American Geophysical Union, Stanford University, December 27, 1962.

4. C. S. Wright, *Scientific American,* September 1962.

5. T. N. Gautier, *Scientific American,* September 1955.

6. New Zealand Department of Scientific and Industrial Research, *Bulletin* No. 40.

7. Report of the Bartol Research Foundation.

8. According to Joseph W. Chamberlain, associate director for space astronomy, Kitt Peak National Observatory, in *Science,* November 1963.

CHAPTER 9

1. Dr. D. Wilwyn John, 1934 *Discovery II* (British) expedition.

2. Quoted in V. Lebedev, *Antarctica,* 1959.

3. Philip, Duke of Edinburgh, *Seabirds in Southern Waters,* 1963.

4. Richard L. Penney in *Natural History,* October 1962.

5. L. E. Richdale, *Sexual Behavior in the Penguin.*

6. Philip, Duke of Edinburgh, *op. cit.*

7. *Natural History Report, Terra Nova* expedition, 1914.

8. Lebedev, *op. cit.*

9. *Ibid.*

10. New Zealand Antarctic Society, *Antarctica Today.*

11. R. C. Murphy, Address to Antarctic Research Program Symposium, 1961.

12. Lebedev, *op. cit.*

13. News Release, National Science Foundation, December 1963.

CHAPTER 10

1. Charles J. Egan and Eugene Evonuk, Arctic Aeromedical Laboratory, Fort Wainwright, Alaska, "Report on Retention of Resistance to Cooling by Alaskan Natives."

2. F. A. Milan, Aeromedical Laboratory, Fort Wainwright, Alaska, *Thermal Stress in the Antarctic.*

3. Neal E. Miller, Yale University, paper read at Centennial meeting, National Academy of Sciences.

4. G. J. Dufek, "What We've Accomplished in Antarctica," *National Geographic,* October 1959.

5. John H. Rohrer, "Human Adjustment to Antarctic Isolation," *Naval Research Reviews,* June 1959.

6. *Ibid.*

7. Charles S. Mullin, Jr., Capt., MC, USN, and H. J. M. Connery, Lt. MSC, USN, "Psychological Study at an Antarctic IGY Station," *Armed Forces Medical Journal,* Vol. 10, No. 3.

8. J. E. Nardini, R. S. Herrmann, and J. E. Rasmussen, USN, report to the American Psychiatric Association meeting May 12, 1961, Chicago, Ill.

9. Reference Division, Central Office of Information, London.

INDEX

Acanthodrilid, 136
Adams, President John Quincy, 14
Adélie Land, 25, 37
Africa, 6, 126, 127, 135, 137, 138, 143, 144, 154, 228, 230
Ahlmann, H. W., 190
Airborne soundings, 87–89
Airglow phenomena, 199, 200, 222–23
Alaska, 58, 154, 156, 212
Albatross, 237
Alexander I, Czar, 13
Alexander I Island, 14, 23
Alexander VI, Pope, 9, 283
Alexandra, Queen, 27
Alexandria, 4, 5
Algae, 227
Algae fossils, 158–59
Alps, 127, 190
Amundsen, Roald, 23, 40, 42–43, 44, 47, 57, 67, 76, 85, 97, 98, 104, 112, 118, 176, 184, 259
Amundsen Sea, 122
Amundsen-Scott Station, 71, 87, 120, 174, 176, 191, 195, 196, 200, 206, 210, 215, 221, 222, 272, 276, 277
Anderson, Vernon H., 74, 76, 78, 98–103, 107, 283
Andes Mountains, 58, 98, 103, 111, 118, 137, 138, 286; *see also* Antarctandes Mountains
Andromache (British brig), 12
Annawan (American ship), 14, 15, 25
Antarctandes Mountains, 58, 103, 111, 121, 123, 137, 138; *see also* Sentinel Mountains
Antarctic (Norwegian ship), 23
Antarctica, African quadrant of, 57, 61, 73; age and formation of, 48–49, 53–54, 58, 129–31, 132–35, 158–60; American quadrant of, 37; Australian quadrant of, 8, 71, 173, 191; animal and plant life of, 54, 66, 100–101, 108, 224–54; annual precipitation in, 169, 196, 260; climatic history of, 48–49, 54, 114, 125, 128, 132–35, 145, 146–147, 153, 157–61, 165, 182–85, 187–189; communication problems in, 107, 108, 178, 181, 200, 201–202, 273–75, 282; in continental drift theory, 127–129, 132, 134–35, 137–38, 143, 228; early speculations about, 3–9, 24, 54; fossil evidence in, 48–49, 54, 125, 127, 129–31, 132–35, 137, 158–60, 228, 249, 252–54; geological structure of, 25–26, 54, 102–103, 107, 282; glaciers and glaciation of, 30, 44ff., 54, 71, 90, 92, 124, 130, 157–61, 162, 164, 165, 168, 169, 176, 182, 187, 198, 255–56; human adaptation to, 255–81; ice thickness determined, 181–82; international

claims upon, 282–86; marine life of, 225–27, 228, 238–54; mineral resources of, 48–49, 125, 129, 130ff., 134, 282; temperature changes in, 191–94, 196–198, 225–26; the search for, 9–14; topography of, 57–58, 60, 63, 73, 76–81, 83, 87, 92, 95–104, 106–107, 110, 112–18, 121–24, 180, 183, 191; weather patterns in, 62, 78, 169–71, 180–83, 187–89, 191–98; a womanless world, 265–68, 278; women visitors to, 62, 267–68
Antarcticans, 24, 256–81
Antarctic Circle, 10, 25, 36, 178
Antarctic convergence, 224–25, 227, 250, 280
Antarctic ice cap 5, 10, 11, 25, 27, 31, 47, 49, 52, 63, 64, 95, 147, 282; age and history of, 143, 144–47, 153, 157–165, 167, 169, 182–85, 187–89, 198, 254; growth of, 140, 148, 150, 158–59, 169, 181, 182–83, 187; recession of, 27, 35, 52–53, 83, 89, 103, 107, 148, 157, 158–61, 169, 182–83; speculation about, 52–53, 58–61, 62, 63, 81, 82, 160; thickness studied, 76–81, 85, 87, 89–93, 96, 98, 100, 103, 181–82, 183; traversed, 56–57
Antarctic islands, 181, 238, 239, 250
Antarctic Ocean, 5, 15, 19, 25, 225, 252
Antarctic Panhandle, 10, 12, 13ff., 23, 24, 37, 57, 58, 61, 62, 98, 101, 103, 104, 117, 118, 122–23, 129, 137, 160–161, 181, 182, 195, 225, 227, 228, 238, 252, 267, 283
Antarctic Peninsula. *See* Antarctic Panhandle
Antarctic Treaty of 1959, 281–82, 284–286
Antarktikos, 3, 4, 286
Antichthon, 3, 4, 5, 7, 286
Appalachian Mountains, 127, 135, 169
Appleton, Edward V., 214
Arctic basin, 18, 19, 40, 59, 62, 131, 185, 189, 192–93, 195, 198, 222, 225, 244, 255
Arctic Circle, 4, 193
Arctic Institute of North America, 74, 88, 222
Arctic Ocean, 59, 63, 139–40, 189, 200
Arctowski, Henryk, 58
Argentina, 62, 63, 74, 120, 172, 195, 218, 242, 244–45, 282, 283, 284
Ariel satellite, 208
Aristotle, 4
Armitage, Lieutenant Albert B., 30
Army-Navy Highway, 71, 73, 75
Asthenosphere, 141
Astrolabe (French ship), 18
Atka (US icebreaker), 65, 67

Atkinson, E. L., 48
Atlantic Ocean, 9, 56ff., 71, 76, 92, 97, 107, 126, 130, 135, 139, 140, 141–43, 147, 181, 196, 202, 207, 214, 221, 224, 237, 238
Atmospherics, 208–13, 218
Aughenbaugh, Nolan B., 106, 108
Aurora, 63, 83, 99, 200, 202, 203–207, 208, 209–10, 213, 216, 222, 279; artificial, 206–207, 208; hissing, 209–10; in IGY research, 200, 203, 206, 207; relation to ionosphere, 216
Australasia, 129
Australia, 16, 53, 59, 60, 63, 127ff., 135ff., 143, 144, 154, 172, 173, 178, 225, 228, 230, 247, 282, 284, 285
Axel Heiberg Glacier, 258
Azores, 284

Bacon, Francis, 126
Balchen, Bernt, 57
Barkhausen, Professor Heinrich, 210–11
Barne, Lieutenant Michael, 31
Barnett, M. A. F., 214
Bartol Research Foundation, 219–20
Basalt, 26, 138
Bathydraconidae, 250
Bausch & Lomb G-15 glasses, 262n.
Bay of Biscay, 244
Bay of Whales, 42ff., 57, 67, 185
Beardmore Glacier, 37, 44, 47, 48, 54, 71, 90, 124, 130, 165, 176
Beatus of Liebana, 5
Beherndt, John C., 106
Belgica (Belgian ship), 23, 40, 58, 129
Belgium, 63, 282
Bellingshausen Sea, 101, 117, 122–23
Bentley, Dr. Charles R., 74, 76, 78–82, 98–103, 107, 108, 113–15, 117, 118, 120ff., 124, 130, 131, 138, 283
Bentley-Anderson traverse (I), 73–81, 85, 96
Bentley-Anderson traverse (II), 98–103, 138, 236
Bentley-Long traverse. See Horlick Mountains traverse
Bering Strait, 40
Berkner, Dr. Lloyd V., 62, 82, 107
Berkner Island, 107, 111
Bertoglio, Lloyd M., 174
Bishop, Captain Ted, 174, 176ff.
Bjaaland, Olav, 43
Blue Glacier, 155
Boothia Peninsula, 8
Borchgrevink, Carsten E., 23–24, 26, 27, 96, 157–58
Bougainville, Chevalier de, 10
Bowers, Henry R., 44, 47ff.
Brachiosaurus, 246
Bradley, Reverend Father Edwin A., S.J., 110, 111–12, 113, 138
Bransfield, Edward, 11, 12
Bransfield Straits, 252
Brasilio Regio, 8
Brazil, 126, 135, 136, 193, 283
Breit, Gregory, 214
Briand, Aristide, 38

Britannia (the royal yacht), 225
Britannia Range, 176
British Admiralty, 10, 12, 31, 61
British Trans-Antarctic Expedition of 1957, 85–87, 105, 117
Brown, R. N. Rudmose, 97
Bruce, William S., 24, 86, 238
Budd Coast. See Wilkes Land
Buenos Aires, Argentina, 40, 169, 218
Buffon, Georges, 126
Buinitsky, V. K., 225
Bull, H. J., 23
Byrd, Rear Admiral Richard E., 57, 62, 64, 65, 67, 71, 72, 76, 97, 98, 112, 113, 130, 184, 185, 259
Byrd Station 71, 73, 75, 80ff., 97, 98, 100, 103, 110ff., 116, 120, 122, 124, 130, 138, 146, 147, 152, 155, 156, 196, 197, 213, 215, 221, 222, 268, 272, 276; after the IGY, 120

C-124 Globemaster aircraft, 71, 174, 176, 177, 179, 279
C-130 Hercules aircraft, 84, 124, 179, 279
Cadiz. See Gades
Calcite rocks, 25, 54, 130, 161, 162, 164, 165
Caledonian Mountains, 127
California, 72, 73, 136, 178, 207, 248
California Institute of Technology, Division of Geological Sciences, 154
Callippos of Cyzicos, 3
Cambrian Period, 144, 148, 165, 167
Cambridge University, 211
Campbell Island, 217
Canada, 18, 137, 189, 191, 202, 203
Canadian Defence Research Board, 213
Cape Adare (Cape Hallett), 19, 20, 23, 26, 27, 35, 66, 85, 158, 179, 207, 218, 227, 286
Cape Armitage, 28
Cape Crozier, 28
Cape Derision, 37
Cape Evans, 40, 44, 47, 193
Cape of Good Hope, 6
Cape Horn, 12
Cape Royds, 36, 230, 235, 236, 238
Cape San Rogue, 126
Cape Verde Islands, 283
Carbon-14, 148–50, 158, 161, 254
Carboniferous Period, 134, 144
Carlson, Senator Frank, 285–86
Carlson, Ronald, 112, 120–21
Carnegie Institute of Washington, Department of Terrestrial Magnetism, 214
Cartographers, ancient, 4–5; American, 12, 13, 124; British, 13; European, 6–9, 12
Cathay, 5, 97
Cenozoic Era, 137
Cesium-137, 187
Chaenichthyidae, 250
Challenger, HMS, 25
Chamberlain, Joseph W., 216
Chang, Fien, 114

Chapman, Sydney, 62
Chapman, William H., 113, 114
Charcot, Jean B., 37–38
Charcot Station, 92
Chile, 14, 63, 128, 129n., 282, 283, 284
Chiloé Island, 128
Christchurch, New Zealand, 27, 33, 147, 178, 179, 217
Christensen, Lars, 61
Christofilos, N. C., 207
Chromium, 107
Cicero, 5
Clams, 252, 254
Clark, John, 189–90
Claudius Ptolemaeus of Alexandria. See Ptolemy
Coal, 48–49, 125, 129ff., 134ff.
Coats Land, 24
Cobalt-60, 164
Cod, 189, 250
Colbeck, William, 30, 157, 158
Coleman, A. P., 60
Coleridge, Samuel Taylor, 20, 237
Colobanthus, 225
Columbia University, 74, 78, 92, 138, 158
Columbus, Christopher, 5, 18, 95, 244
Comité Spécial de l'Année Géophysique Internationale (CSAGI), 63
Commonwealth Bay, 37
Compania Argentina de Pesca Station, 244–45
Constantinople, 8
Continental drift theories, 125–29, 132, 134–38, 140–43, 150; opponents of, 138–40, 143
Cook, Frederick, 23
Cook, Captain James, 10, 11, 27, 170
Cook, John C., 87
Corals, 252, 254
Cosmic rays, 200, 218–22
Cowes, England, 27
Craddock, Campbell, 120
Crary, Albert P., 76, 82, 89–91, 92, 116–117, 118, 119, 122, 253, 279
Crary Range, 99
Crater Hill, 28
Cretaceous Period, 137, 143, 144, 169
Crozier, Francis M., 22
Crustaceans, 136, 249, 250
Cruzen, Richard H., 62

Dadoxylon, 130
Dalrymple, Alexander, 10
Darby, David G., 252, 254
Darlington, Jennie, 62, 267
Dartmouth College, 211
Darwin, Charles, 136–37
David, T. W. Edgeworth, 37, 60
Dawson, Merle (Skip), 71
Debenham, Frank, 54, 252–53, 254
Deception Island, 12, 284
De Gerlache, Adrien, 23
De Havilland "Otter" aircraft, 106
Dehydration, 261–62
De Paul University, 251
Deschampsia, 224

Deutschland (German ship), 56
Devonian Period, 132–34, 136, 144
Dietz, Robert S., 143
Diplodocus, 246n.
Discovery expedition, 26–35, 37, 39, 40, 49, 276; subsequent expeditions, 61
Discovery II (British ship), 61
Dissostichus mawsoni, 250, 251
Dolphins, 40, 247
Donn, William L., 139–40, 145
Dorf, Erling, 189–90
Doumani, George A., 114, 132–34
Drake, Sir Francis, 9, 238
Drake Passage, 58, 195, 227
Dufek, George J., 65, 70, 71, 107, 120, 267
Dufek Massif, 107–108
Duke of Sussex, 214
Dumont d'Urville, Jules Sebastian César, 18, 22, 25, 228
Dundee Island, 57
Du Toit, Alexander L., 137

Earth, atmosphere of, 199–203, 207; convection currents within, 141–42; geophysical history of, 96, 139–150; water budget of, 92–93
Earthworms, 135, 136, 249
East Antarctica, 16, 60, 71, 85, 110, 120, 137, 170, 221; topography of, 92, 96, 111, 113, 116–17, 124, 180, 183, 191; weather patterns in, 170, 180–83, 191, 196
East Antarctic ice cap, 62, 87–93, 181–183
Edisto, USS, 166, 185
Edward VII, King, 27
Eights, James, 15, 25
Eights Coast, 101, 123, 286
Eights Station, 196, 222
Eisenhower, President Dwight D., 118
Eklund, Carl R., 225, 228, 234
Electromagnetic radiation, 199–213, 223; measurement of, 217–18
Elephant Island, 56
Elizabeth I, Queen, 9
Ellsworth, Lincoln, 57, 60, 76, 78
Ellsworth Highland, 14, 107, 122, 181, 222
Ellsworth Land, 104, 137, 196
Ellsworth Station, 71, 104, 105, 106, 108, 110, 111, 138, 206, 215; assigned to Argentina, 120
Ellsworth traverse, 104–10, 236
Eltanin (US research ship), 252, 267
Enderby Land, 61, 137, 247
Endurance (British ship), 56
Engle, Senator Clair, 284–85
Enniskillen Dragoons, 44
Eocene Age, 144, 228
Epperly, Robert, 115
Epstein, Samuel, 154
Eratosthenes, 4
Eric the Red, 190
Erebus (British ship), 18, 20, 22
Eskimos, 38, 189, 255, 256, 262
Eucryphia, 129

Eudoxus of Cnidos, 3
Europe, 6, 8, 12, 17, 59, 126ff., 137, 139, 144, 190, 202, 225, 239, 240, 245, 284
Evans, Edgar, 30, 44, 48, 49
Evteev, Sven, 279
Ewing, W. Maurice, 138–40, 145, 168
Explosion seismology, 72, 78–81, 85, 87, 90, 97

Falkland Islands Dependencies, 61, 284
Fanning, Edmund, 14
Ferrar, Hartley T., 30n., 130
Ferrar Glacier, 30
Filchner, Wilhelm, 56, 86, 97, 104
Filchner channel, 114, 117, 122; see also Great Antarctic Trough
Filchner Ice Shelf, 66, 71, 87, 104–10, 111, 122, 130, 236
Filchner traverse. See Ellsworth traverse
Finland, 144, 189
Flint, Richard Foster, 60
Florian (Florianus), Antonio, 7, 8
Flowers, Edwin C., 195
Flying Fish (US tender), 15, 16
Fort Churchill, Manitoba, 203
Fossils, 48–49, 54, 125, 128, 129–32, 134ff., 158ff., 228, 249, 252–54
Fram (Norwegian polar vessel), 40, 42, 44, 89
Framheim, 42, 43, 44, 184, 185
France, 10, 14, 22, 63, 89, 104, 135, 172, 282
Fremow, E. J., 206
Frostbite, 260, 263–64, 265
Fryxell Glaciation, 158
Fuchs, Sir Vivian, 85–87, 92, 104, 105, 117, 119
Fulbright, Senator William J., 285

Gades (Cadiz), 4, 5
Gallirallus australis hectori, 75
Gastropods, 129
Gauss, Johann Karl Friedrich, 18, 214
Georgetown University, 92
Georgetown University Medical School, 273
George Washington University, 280
Germany, 38, 135
German High Command, 210
Gilbert, William, 18, 214
Giovinetto, Mario B., 74, 78
Glacial Lake Washburn, 158
Glacier, USS, 65, 66, 146, 147–48
Glacierization, 52, 59, 139–40, 143–44, 156–61, 169, 187, 190, 198, 256
Glossopteris, 48, 125, 128, 129, 130, 134
Gneiss Point, 165
Gneisses, 25, 54
Gobi desert, 54
Godthaab, Greenland, 136
Golden Hind (British ship), 9
Gondwanaland, 127–28, 132, 135, 136ff., 161, 228
Goodwin, A., 111
Gould, Laurence M., 60, 64, 130–31, 259
Gow, Anthony J., 148, 153, 155, 254
Graham, Sir James R. G., 13

Graham Land, 13, 58, 61; see also Antarctic Panhandle
Grand Chasm, 106
Gravity soundings, 78–79, 96
Great Antarctic Horst. See Trans-Antarctic Mountains
Great Antarctic Trough, 57, 73, 76, 96, 97, 98–103, 106–107, 110, 111–15, 117, 118, 121–23
Great Britain, 4, 9, 10, 11, 24, 26, 27, 30, 31, 38, 47, 56, 63, 127, 135, 136, 173, 190, 208, 211, 214, 225, 238, 282, 283, 284; government of, 283, 284
Great Whale River, 213
Great Whale Station, 213
Greenland, 23, 43, 47, 71, 127, 131, 136, 140, 150, 152, 154, 187, 190, 220, 244
Greenland ice cap, 59, 60, 73, 156, 167, 187, 188, 191
Greenwich Mean Time, 173
Grenville, Sir Richard, 9

Hansen, Helmer, 43
Harlin, Ben W., 196–98
Hartog, Stephen Den, 116
Hassel, Sverre, 43
Hatherton, Trevor, 116
Hawkes, William M. (Trigger), 107
Hale, Daniel P., 98, 103
Hallett Station, 66, 120, 195, 196, 207, 218, 227
Halley Bay Station, 221, 222
Heaviside, Oliver, 214
Hedblom, Earland E. (Doc), 185, 261ff., 265
Heezen, Bruce C., 140, 145
Helium, 201
Hemiptera (Peloridiidae), 129
Henderson, Sidney, 177
Henson, Matthew, 38
Hepinstall, Patricia, 267
Hero (American sloop), 12
High Jump, Operation, 62, 63
Hillary, Sid Edmund, 85, 87
Himalayas, 143
Hitler, Adolf, 57, 284
Hobbs, William H., 12
Holdgate, Martin, 128–29
Hollow-earth theory, 17
Hooker, Joseph, 128, 136
Horlick, William, 113
Horlick Mountains, 111, 113–15, 117, 121–22, 125, 131, 132–35, 165, 182
Horlick Mountains traverse, 111, 113–15, 120, 130, 131
Horst regions, 138
Hudson Bay, 213
Hui-te-rangiora, 5
Huronian Era, 144
Hut Point, 28, 29ff., 36, 37, 40, 66, 276
Hydrogen, 200, 201
Hydroxyl molecule, 222

Ice age theories, 52–53, 59–60, 72, 81, 93, 96, 132, 139–40, 143, 145, 156, 159, 160, 165, 168, 183–85, 187, 189, 198

Ice core dating, 150–55
Iceland, 38, 127, 189, 190, 244
India, 5, 95, 125, 127, 129, 135, 137, 143, 144
Indian Ocean, 6, 7, 10, 120, 139, 140, 191, 247, 250
Injun I, radiation-detector satellite, 208
Insects, wingless, 54, 224
Institute of Earth Sciences (University of Toronto), 141
Institute of Polar Studies (Ohio State University), 132
International Council of Scientific Unions (ICSU), 63
International Court of Justice, 284
International Geophysical Cooperation (IGC), regime of, 118; Argentinian contribution to, 120, 227, 282; Australian contribution to, 282; Belgian participation in, 282; British role in, 217, 220, 221-22, 282; New Zealand contribution to, 217–18, 279, 282; Russian contribution to, 217, 220, 221–222, 279, 282, 283, 285; U.S. Antarctic role in, 119–24, 132–34, 161–65, 209–10, 212, 217, 220, 221–22, 250–254, 258–59, 282
International Geophysical Year (IGY), Antarctic Research Program, 56, 283, 284; Antarctic explorations, 73, 76–81, 85–87, 89–92, 96, 98–103, 106–10, 112–18; aurora research, 200, 204, 206, 207; British contribution to, 173, 283; ice thickness probes, 73, 76–81, 85–87, 89–93, 96, 98, 100, 103; New Zealand participation in, 148, 155, 172, 207, 217–18; objectives of, 62–64, 73, 82–83, 119; participants, 63–64, 72, 173, 181, 213; physiological adaptation research, 259–60; preparations for, 64–71, 119, 184; Russian contribution to, 172, 181, 191, 283–84, 285; study of Antarctic climatic history, 146–54, 156, 158, 159–60; study of Antarctic meteorological profile, 170–173, 182, 191–92, 194–98; study of upper Antarctic atmosphere, 199, 200–201, 204, 206, 207, 208–209, 215–18, 220; topographical studies, 98–103, 106–107, 110, 112–18; U.S. role in, 62–71, 73–83, 87–92, 96, 98–117, 130–131, 138, 146–55, 156, 158, 159–60, 172–73, 182, 184, 194–98, 207, 218, 234, 259; zoological experiments, 234, 236
International Union of Geodesy and Geophysics, 62
Ionosphere, 63, 178, 179, 199, 201–203, 207, 209, 210ff., 214–18, 220–22, 273; regions of, 215–16, 220–21; soundings of, 217–18
Iron, 107
Isabella, Queen, 5
Isostatic readjustment, 144, 160

Jackson, Senator Henry M., 285
Jacobus Angelus, 7

Japan, 63, 245, 282
Jamesway huts, 66, 71, 174
Jason (Norwegian ship), 129
JATO bottles, 70, 100
Jellyfish, 250
Johns Hopkins University, 100
Johnston Island, 206–207, 208
Jones, Marshall B., 280
Jones, Thomas O., 119
June, H., 57
Jurassic Period, 144, 188

Kainan Bay, 67, 73, 185
KC-4-USV radio station, 273
Kelly, Ruth, 267
Kennelly, Arthur E., 214
Kennelly-Heaviside Layer, 214; *see also* Ionosphere
Kennedy, President John F., 179
Kentucky blue grass, 227
Kenyon, H. Hollick, 57
Kerguelin-Tremarec, Yves Joseph de, 10
Kerguelin Island, 10, 225
Kernlose winter, 193
King Edward VII Land, 28, 35, 37, 137
King Haakon VII's Plateau, 43; *see also* South Polar Plateau
Kirwan, L. P., 19
Kitt Peak National Observatory, 222
Koettlitz Glaciation, 158–59
Krill (*Euphausia superba*), 227, 246, 251–52
Kristensen, Leonard, 23
Kukri Hills, 30

Lamont Geological Observatory, 92, 138, 140, 158–59
Lanterman, William S., 174, 178
Larsen, C. A., 23, 129, 244–45
Laurelia, 129
Lawrence Radiation Laboratory, 207
Layman, Frank, 116
Lazarev Station, 279
LeSchack, Leonard, 113
Levick, G. Murray, 230, 233
Lichens, 54, 103, 224, 227–28
Lightning, 211–12
Limestones, 25, 54, 130, 161
Lithium, 207
Little America, 57, 184, 185–87
Little America I, 184, 185
Little America II, 184, 185
Little America III, 184, 185, 186
Little America V, 67, 71, 73, 74, 76, 80ff., 89, 90, 97, 155, 172, 173, 184, 185, 192, 194, 196, 206, 207, 215, 259, 260; closed, 115–16, 185
Lobsters, 250
Long, Jack, 98, 100, 103, 113, 114, 130
Long, William E., 98, 103, 113, 114, 130, 132–34
Lord Howe Island, 129
Louis Philippe, King of France, 18

Macmillan, Harold, 285
Macquarie Islands, 212, 225
Mac-Robertson Land, 286

Macrobius, 5
Madagascar, 127, 135, 137
Magellan, Ferdinand, 7
Magnetic field phenomena, 17–18, 78, 85, 199–218
Magnetic soundings, 78, 96
Magnetic storms, 107, 108, 201–203, 215; artificial, 206–208
Malachite (copper carbonate), 107
Manchurian ponies, 37
Marble Point, 162, 282
Marciniak, Phyllis L., 251–52
Marconi, Guglielmo, 214
Marianas Islands, 8
Marie Byrd Land, 16, 60, 62, 67, 71, 74, 76, 82, 89, 90, 97, 108, 110, 112, 113, 122, 138, 146, 148, 150, 152, 155, 156, 180, 181, 253, 283, 285; topography of, 76–81, 98–103, 111, 112–13, 117
Marquesas, 9
Marshall, Earnest W., 148, 153–54, 155, 156
Maudheim, 196
Mawson, Sir Douglas, 76, 83, 96, 97
McDonald, Edwin A., 108
McGinnis, Lyle, 116
McKinley, A. C., 57
McMurdo Glaciation, 158
McMurdo Sound, 20, 24, 28, 31, 35, 36, 40, 44, 53, 58, 61, 65, 67, 70, 71, 87, 90, 92, 115, 121, 128, 129, 136, 145, 146, 156ff., 159ff., 173, 178, 179, 182, 183, 194, 206, 220ff., 230, 233, 241, 242, 248–49, 250, 251, 253, 264, 267, 268, 282; region examined, 156–60, 162–65, 166
McMurdo Station. See Naval Air Facility McMurdo
McWhinnie, Dr. Mary A., 251–52
Melbourne, Australia, 40, 173
Melville, Herman, 247
Mercator, Gerhard (né Kremer), 8
Mesosaurus, 135, 136
Mesozoic Era, 58, 134, 137, 142, 246
Mica-schists, 25, 26
Midwinter Day, 275, 276, 279
Mirnyi (Russian ship), 13
Mirnyi Station, 279
Moby Dick, 247, 248
Mollusks, 129, 130, 132, 227, 247, 249, 250, 252, 254
Moraines, 27, 35, 107
Moreland, W. B., 172–73, 180–81
Morency, Anthony J., 74, 75
Morgan, M. G., 211
Morning (British relief ship), 29–30, 31, 36
Morozumi, Henry M., 209–10
Mount Betty, 258
Mount Erebus, 20, 21, 28, 29, 37, 40, 66, 77, 145, 154, 166, 254, 277, 282
Mount Glossopteris, 130
Mount Katmai, 154
Mount Krakatoa, 152, 154
Mount Nussbaum, 160
Mount Peterson, 124
Mount Takahe, 99, 100, 102

Mount Terror, 20
Mount Tuve, 124
Mulock, George F. A., 31
Murphy, Robert Cushman, 245
Murray, John, 25–26, 49, 96, 117, 118, 170, 181
Murray's Hypothetical Continent, 49, 96–97, 117, 170, 181

Nansen, Fridtjof, 40, 189
National Academy of Sciences, 82
National Bureau of Standards (NBS), 221–22
National Geographic Society, 124
National Science Foundation, 119, 252, 267; Office of Antarctic Research, 119; Office of Special International Programs, 119
Naval Air Facility McMurdo (NAF McMurdo), 65, 66–69, 87, 88, 91, 116, 160, 173, 184, 185, 217, 221, 264, 267, 272, 273, 276ff., 282; after the IGY, 120
Neuburg, Hugo A. C., 104, 105–10, 111, 112–13
Newfoundland, 127, 244
New Hebrides, 9
Newnes, Sir George, 23
New York Academy of Science, 15
New York Aquarium, 241
New York Lyceum of Natural History. See New York Academy of Science
New York University, 104
New Zealand, 27, 31, 36, 56, 59, 63, 64, 75, 85, 92, 99, 108, 120, 128–29, 135, 136, 137, 147, 172, 174, 178, 179, 206, 212, 217, 218, 225, 228, 230, 231, 236n., 237, 245, 247, 254, 267, 279, 282, 284, 285; North Island of, 206; penguins of, 231–32, 233; South Island of, 27, 179, 206, 217
Niagara Falls, 144
Nichols, Robert L., 60, 62, 160–61
Nimrod (British whaling ship), 36
Nitrogen, 201
Nordenskjold, Otto, 24, 244
Norse colonies, 190
North America, 10, 54, 59, 108, 126, 127, 139, 144, 167, 168, 183
North Atlantic Ocean, 92, 135, 140, 147, 207, 237; see also Atlantic Ocean
North pole (geographic), 17, 38, 59, 63, 132, 143, 283
Northern hemisphere, 59, 60, 118, 135, 139, 140, 144, 158ff., 167, 183, 212, 228, 244; climatic change in, 189–91, 193, 195
Norway, 63, 127, 135, 140, 184, 189, 190, 244, 282
Nothofagus, 129
Notothenia rossi, 250
Nototheniidae, 250, 252
Nunataks, 98–99, 103, 121, 227

Oates, Lawrence E. G., 44, 46, 48, 49
Oates Land, 16
Observation Hill, 49

Octopuses, 250; *see also* Squids
Odishaw, Hugh, 82
O'Higgins, Bernardo, 14
O'Higgins Land, 13–14
Ohio Range (of Horlick Mts.), 132–35
Ohio State University, 98, 106, 119, 130, 132, 227
Oikoumene, 286
Olympia National Park, 155
Operation Deep Freeze, 65, 67, 176, 269
Orbis Imago (1538), 8
Ordovician Period, 144
Oronce Fine (Orontius), 8
Ortelius, Abraham, of Antwerp, 8
Ostenso, Ned A., 74, 78–80, 82, 98, 103, 111, 112–13, 117ff.
Oxygen, 194, 201, 216, 222, 223, 226, 251
Oxygen-18, 154–55

Pacific Naval Laboratory of Canada, 213
Pacific Ocean, 16, 25, 27, 40, 56, 90, 92, 97, 112, 129, 130, 137ff., 141, 143, 147, 178, 181, 207, 230
Paleozoic Era, 134, 137, 144
Palmer, Nathaniel Brown, 12–13, 14, 238
Palmer Peninsula, 13
Palmer's Land, 13
Pangaea, 127, 128, 132, 135–37, 142
Patagonia, 8, 58, 137, 225, 228
Peacock (American sloop), 15
Peary, Robert E., 17, 38
Pelorididiidae, 129
Pendleton, Benjamin, 14
Penguins, Adélie, 36, 101, 108, 228, 230–31, 232, 234, 235–36, 241; Antarctic (Spheniscidae), 55, 66, 100–101, 108, 225, 228–36, 237, 238, 241; Chinstrap, 230; Emperor (*Aptenodytes forsteri*), 108, 227, 228, 230, 234–35, 236; Gentoo, 230, 232–33; King, 228, 230; New Zealand, 231–32, 233
Penney, Richard L., 230
Pensacola Mountains, 106, 107–108, 182
Pensacola Naval Air Station, 107
Permian Period, 131, 134, 136, 144
Petrels, 227, 237
Péwé, Troy L., 156, 158, 159–60, 165
Philip II of Spain, 9
Phleger, Herman, 286
Phormidium incrustatum, 108
Pickering, Edward C., 126
Pirritt, John, 110, 111–12, 113, 138
Pirritt-Bradley traverse, 111–12, 113, 138
Pitchblende, 153
Plankton, 55, 226–27, 228
Pleistocene Epoch, 58, 60, 139, 140, 143, 156, 158, 159, 167, 190, 256
Podocarpus (Black pine), 129
Pole of Antarctic wind circulation, 182
Pole of relative inaccessibility, 64, 182
Pole Station. *See* Amundsen-Scott Station
Polynesia, 18
Pomponius Mela, 5
Ponting, Herbert G., 21, 34, 39, 40, 41, 42, 46

Porpoise (American brig), 15, 18
Porpoises, 247
Port Lyttleton, 27, 33, 147
Portugal, 9, 283
Posidonius, 4
Potassium-40, 141
Poulter, Thomas C., 72
Pourquoi Pas? (French ship), 37–38
Pre-Cambrian times, 54, 58, 103, 111, 121, 144
Priestley, Sir Raymond, 52, 57, 58, 96, 129
Prince Albert Mountains, 129
Prince Philip of England, 225, 233
Princess Martha Land, 137
Princess Ragnhild Land, 137
Project Apollo, 38
Project Argus, 207–208
Project Mercury, 203, 281
Ptolemaic maps, 7
Ptolemy, 5, 6
Purdue University, 106
Pyroheliometers, 195
Pythagoras, 3
Pytheas of Massilia (Marseilles), 4

Quartziferous diorites, 25
Quartzite, 54
Quaternary Period, 158
Queen Alexandra Mountains, 176
Queen Maud Land, 57, 61, 73
Queen Maud Mountains, 130, 176, 258
Que Sera Sera (Navy R4D aircraft), 70

Radiocarbon dating. *See* Carbon-14
Radiosonde balloons, 201
Rarotonga, 5, 217
Ray, Carleton, 242
Relief (American storage ship), 15
Revelle, Roger, 151–52, 153, 154
Richdale, L. E., 231–32, 233
Rime of the Ancient Mariner, 237
Ritscher, Alfred, 57
Rivard, Norman R., 156, 158, 159–60, 165
Robert English Coast, 123
Robinson, Edwin, 120
Rocky Mountains, 58, 137
Rohrer, John H., 273, 275, 279
Rolligon cargo platform, 92
Ronca, Luciana, 165
Ronne, Edith, 62, 267
Ronne, Finn, 62, 104, 108, 124, 267
Roosevelt Island, 89
Ross, James Clark, 16, 18–22, 25, 27, 28, 49, 52, 61, 117, 128, 157, 214, 244
Ross Ice Shelf, 22, 24, 27, 28ff., 35, 57, 67, 72, 73, 97, 100, 102, 104, 110, 112, 116, 118, 130, 155, 176, 177, 185, 194, 195, 240, 252, 259; composition and thickness of, 89–90, 155, 157–58, 252–254; movement of, 42, 43, 44, 52–53, 58; sources of, 89–90, 252–54
Ross Ice Shelf traverse (Crary), 89–90
Ross Island, 20, 27, 28, 31, 36, 37, 40, 66, 87, 129, 145, 157, 174, 184, 217, 238, 250, 264, 277

Ross Sea, 19, 23, 24, 27, 31, 41, 44, 53, 56, 57, 58, 60, 65, 66, 73, 86, 89, 96, 103, 107, 112, 113, 126, 137, 146, 154, 155, 157, 158, 180, 181, 184, 185, 217, 227, 240, 253
Ross Sea Dependency, 64, 218, 282
Royal Society of London, 25, 117, 214
Royal Society Mountains, 54, 157
Royal Navy, 12, 26, 157
Royds, Charles W. R., 28
Rubin, Morton J., 170, 192–93
Russell, Senator Richard B., 285
Russia, 13, 14, 63, 135, 144, 172, 282, 283
Rymill, John, 61

Sabine Island, 136
Salamanca, Antonio, of Milan, 8
San Andreas fault, 136
Sandstones, 25, 26, 48, 54, 129, 134
Schists, 25, 54
Schoner, Johannes, 8
Scorpions, 135
Scotia Sea, 56
Scott, Robert Falcon, 22, 26–35, 36, 39–40, 42, 52, 53, 58, 59, 66, 70, 71, 83, 85, 96, 118, 129, 130, 132, 136, 138, 157, 169, 170, 176, 187, 193, 204, 230, 265, 276
Scott Base, 217, 218, 279, 282
Scott Polar Research Institute, 128
Scottish National Expedition, 24, 83, 238
Seabees, 64, 67, 71, 104, 116, 222
Sea Gull (New York pilot boat), 15
Seal hunting, 15, 17, 25, 227, 238–40, 242
Seals, 14, 55, 225, 227, 238–42, 249, 250; Crabeater, 160, 227, 240–41; eared, 238, 239; Elephant, 14, 161, 238, 239, 240–41; Leopard, 240, 242; Ross, 241; true, 238, 239, 240; Weddell, 239, 240, 241–42, 243
Sea urchins, 249
Sedov (Russian icebreaker), 189
Seismic soundings, 78–81, 85, 87, 90, 96, 97, 106
Sentinel Mountains, 98, 100, 102–103, 107, 110, 111, 113, 117, 118, 138; connection with Trans-Antarctics sought, 120, 121, 122–24
Sentinel Mountains traverse. See Bentley-Anderson traverse (II)
Seward, A. C., 137
Seymour Island, 228
Shackleton, Ernest Henry, 29, 36–37, 40, 44, 49, 56, 86, 96, 97, 130, 265, 281
Shackleton Station, 87, 104, 105
Shales, 25, 130
Sharp, Robert P., 154
Sheffield, A. H., 173
Shetland Islands, 4n.
Shinn, Conrad, 70
Silurian Period, 144
Simpson, George C., 170, 193
Siniuk, Joseph, 176, 177
Siple, Paul, 71
Skelton, Reginald W., 30

Skelton Glacier, 30, 90, 92
Skuas, 235, 237–38
Sladen, William J. L., 100, 101, 108
Slava (Russian whale-factory ship), 245
Smith, William, 11–12, 238
Smith, William M., 280
Snails, 129, 132, 252
Snowblindness, 262–63
Sodium, 222
Solar-flare warning system, 263
Solar radiation, 59, 63, 83, 85, 132, 139, 187–88, 194, 195, 199–213, 215, 216–218, 223
Solomon Islands, 8
Sorge, Ernst G., 150–51, 152
Sorge's Law, 151, 152
South Africa, 63, 125, 127, 128, 135ff., 172, 247, 282
South America, 8, 24, 37, 56ff., 118, 126, 127, 135, 136, 138, 143, 144, 154, 202, 228, 240, 247, 283
South Antillean arc, 137
South Atlantic Ocean, 58, 126, 207
South Georgia, Island of, 11, 14, 56, 225, 236–37, 244, 245, 250
South Ice base, 87
South Orkney Islands, 23, 61
South Pacific Ocean, 16, 129, 207
South Polar Plateau, 30, 35, 37, 43, 44, 47, 52, 53, 87–89, 90–92, 98, 103, 110ff., 116–17, 125, 138, 165, 170, 174, 176, 177, 187–88, 195, 253
South pole (geographic), 9, 17, 24, 25, 37, 49, 59, 63, 64, 70, 71, 87, 92, 132, 136, 137, 140, 173, 176, 179, 182, 184, 191, 192, 195, 196, 199, 200ff., 209, 210, 215ff., 221, 222, 258ff., 268, 279, 284, 287; ice thickness at, 92; race to reach, 39, 43–48, 137, 176, 184
South pole (magnetic), 18–22, 37, 60, 182, 214
South Sandwich Islands, 283
South Shetland Islands, 12, 14, 15, 16, 56, 61, 250, 283, 284
Southern Cross (British ship), 23, 158
Southern hemisphere, 10, 59, 96, 127ff., 135, 136, 172, 178, 212, 228, 240, 244, 282
Southwestern Research Institute, 87–88
Soviet Union. See Russia
Spain, 4, 5, 9, 53, 89, 283
Spermaceti, 244, 245
Spitsbergen, 131, 184–85, 189, 244
Sponges, 249–50, 252, 254
Sputnik, 85
Squids, 227, 247
Stanford Research Institute, 72
Stanford University, 212, 213, 246, 250, 251
Stanford University Radio Science Laboratory, 209, 212–13
Starfish, 249
Starfish Prime bomb test, 208
State University of Iowa, 208
Staten Island (US icebreaker), 104, 108, 109